T0085979

THE
WEATHER
BOOK

THE
WEATHER
BOOK

WHY IT HAPPENS AND
WHERE IT COMES FROM

DIANA CRAIG

Michael O'Mara Books Limited

This revised paperback edition first published in 2024

First published in Great Britain in 2009 by
Michael O'Mara Books Limited
9 Lion Yard
Tremadoc Road
London SW4 7NQ

A CIP catalogue record for this book is available from
the British Library.

Papers used by Michael O'Mara Books Limited are natural, recyclable
products made from wood grown in sustainable forests. The
manufacturing processes conform to the environmental regulations
of the country of origin.

ISBN: 978-1-78929-590-0 in paperback print format

ISBN: 978-1-84317-843-9 in ebook format

1 2 3 4 5 6 7 8 9 10

www.mombooks.com

Illustrations copyright © Sailesh Patel 2009, 2024

Typeset and designed by e-type

Printed and bound by CPI Group (UK) Ltd, Croydon, CR0 4YY

Contents

Dedication

I should like to dedicate this book to my intrepid daughters, Blanche and Charlotte, who have travelled to the far reaches of the world, from Brazil to Patagonia and Kerala to Alaska, and have seen so much of this amazing planet in all weathers and climates.

Acknowledgements

The weather is a complex subject and trying to explain it all within the pages of a relatively short book can be a challenging prospect. I should like to thank the team at Michael O'Mara for making my job easier, and for all the support they have given me during the writing process. In particular, I should like to thank: Carol Kirkwood for her excellent foreword; Helen Young for lending her expert eye; Toby Buchan for his unfailing courtesy and efficiency; and Ed Pickford for designing the book and bringing the words to life on the page.

Foreword

∽

I was lucky to grow up in one of the most beautiful parts of the United Kingdom, on the West Coast of Scotland. Even as a child, I loved the way the weather could change the landscape within minutes. Dark, scudding, rain-filled clouds would gather over the islands of Eigg and Rhum and then suddenly golden shafts of sunlight would burst across the silver sands of Morar and turn the clouds into gilt-edged candyfloss.

So it's hardly surprising that I grew up with a fascination for the weather and how it affects all of us. I cannot admit to having any desire to be a weather presenter when I was that young, but I will say that from the moment I made my first televised forecast I knew I was absolutely going to love my job. There are very few subjects covered on television that affect everyone – and which everyone has an opinion about.

Our understanding of meteorological forecasting has dramatically improved over the years, as has our method of demonstrating it on screen. When I first started presenting the weather, we still used symbols for sunshine, clouds and rain that a viewer of the 1970s would instantly recognize. Today's graphics are much more realistic and reflect not only our understanding of the weather and how it moves around the UK but also the power of today's computers to create the pictures.

Computers have revolutionized our ability to forecast, and the Met Office uses a supercomputer to predict ever more accurate short- and long-range forecasts, allowing even earlier warning of low-probability, high-impact weather. The system can even help research climate change and its impact on society and the economy. In fact, the Met Office's latest supercomputer is projected to become the most advanced machine in the world dedicated to weather and climate. If nothing else, that shows you how much importance we attach to the weather in this country.

As you know, even with all that computing power and some of the brightest brains in meteorology – and I'm privileged to work with them on a daily basis – we don't always get the weather forecast correct, but you can bet that when it isn't spot on, there's no one more frustrated than a Met Office forecaster!

I hope that if you're reading this book you've developed as much of a fascination with the weather as I have. If I'm correct, then the wealth of information in these carefully researched pages will help you to learn more about the amazing climate that our islands enjoy – and who knows, you might be able to forecast the weather yourself. If you do – and you get it right more often than I do – just don't write and tell me!

Carol Kirkwood

Introduction

MOST OF US check the weather forecast because we want to know how it will affect our day-to-day lives. Will we be able to have that family picnic at the weekend? Will we have to turn on the central heating? Will it rain so that that hosepipe ban will be lifted? When will the hot, humid weather break so we can sleep comfortably at night? That's usually as far as our interest in the weather goes.

But our understanding of the weather and its causes is of vital importance to human survival. While most of us are no longer vulnerable to the extremes of weather that would have killed off many of our ancestors – who did not have the luxury of heated, air-conditioned, insulated housing – global warming and its effects is something that threatens to affect us all. Now more than ever it is important to understand what weather patterns mean, and how we can predict any changes that might have an impact on human life.

The Weather Book helps the reader understand the basic principles of weather formation and explains not only the localized effects of weather but also the global repercussions of climate change, allowing readers to talk with confidence on the topic that is on everybody's lips.

SECTION 1: FORECASTING

1. Red Sky at Night: Early Forms of Forecasting

∽

People have always been fascinated by – and in awe of – the elements, and with good reason. Long ago, before food grown in one country could be flown right across the world to another country, populations were much more dependent on what they produced locally – and hence on the weather that could spell success or ruin for a crop. Way back then, before modern meteorology and hi-tech forecasting, it must have seemed to many that they were at the mercy of capricious forces they did not understand.

Blowing in the wind

Wind vanes are one of the first forms of forecasting equipment – going back to the era BC, in fact, and ancient Babylon, Egypt, China and Greece. The most celebrated of these old-world vanes sat at the top of the Tower of the Winds, a structure 12 metres (approx. 40 feet) high that was built by the astronomer Andonikos in Athens in about 50 BC. Depictions of the gods associated with each of the eight wind directions were

carved into its side, and a bronze Triton holding a rod stood on top of the tower. When the wind blew, he would spin on his axis and point his rod into the wind. Later, wind vanes were a popular addition to the European churches of the Middle Ages.

The sky's the limit

In the fourth century BC, one of the greatest of all Greek philosophers, Aristotle – who was also tutor to Alexander the Great– produced the first serious study of the atmosphere. In this, he presented his theories on a wide variety of natural phenomena, from rainbows to snow. He entitled his work *Meteorologica*, after the Greek word 'meteorol', meaning an occurrence in the sky. From this we get both the words 'meteor' and 'meteorology'– the study of heavenly happenings.

A NUMBERS GAME

According to the Italian explorer Marco Polo (*c.*1254–*c.*1324), the great Mongol conqueror and emperor Kublai Khan (1215–94) kept around 5,000 court astrologers. Among their duties was forecasting the weather – and getting it right or wrong was a matter of life or death. But with so many other astrologers on hand, there was always another to take the place of the one who had 'retired early'.

Let your hair down

It isn't particularly scientific, but you can observe the behaviour of your hair for an indication of how wet the air is. Like paper, hair shrinks when it is dry and stretches when it is moist, increasing by up to 2.5 per cent in length. This effect did not go unnoticed centuries ago and became the basis of a number of methods for recording humidity:

- The German Nicholas of Cusa (*c.*1401–64) invented a method for measuring humidity by checking the amount of moisture absorbed by wool.
- Others constructed hygrometers (humidity-recording devices) using, among other materials, human hair, ox's intestine, rat's bladder, strings and wild oats – all of which responded to a change in the moisture content of the air.

Viva Italia!

Renaissance Italians were ahead of the game when it came to early meteorological inventions:

- In about 1450, the Italian architect Leon Battista Alberti (1404–72) invented the first anemometer, a device for measuring windspeed.
- In about 1592, Galileo Galilei (1564–1642) invented an early thermometer.
- In 1643, Evangelista Torricelli (1608–47) – who was a man, not a woman as the name suggests – invented the mercury barometer for measuring atmospheric pressure.

Nature's language

Uneducated and without either the resources or leisure time to perform scientific experiments, ordinary people once relied on careful observation of nature to predict the weather, and a whole body of weather lore developed. Here are just a few examples…

It will be rainy and stormy if:

- flowers close up.
- the cows are lying down.
- there's a ring of cloud around the moon.
- the swallows fly low (if they are flying high, the winds are light).

It will turn into a fine day if:

- it rains before seven.
- it's foggy in the morning.

Winter will be severe if:

- squirrels store large quantities of nuts.
- squirrels have bushy tails.
- there are lots of berries on the trees.
- ant hills are high in July.
- hornets build their nests high.
- November is warm (obviously this applies only to the Northern Hemisphere, where this is a winter month).

A ROSY GLOW

Not all folklore about the weather is old wives' tales. The well-known saying 'Red sky at night, shepherd's delight; red sky in the morning, shepherd's warning' (or the American version, in which 'shepherd' is 'sailor') may have some foundation in fact. A red sunset sky is usually caused by reflections from minuscule dust particles trapped high in the atmosphere by conditions that occur before fine weather.

2. Weather Watch: Meteorology

∽

For many people, knowing what the weather is going to do affects them no more than choosing whether to take an umbrella to work, or to have that picnic or barbecue at the weekend. But for others – farmers, pilots, sailors – accurate forecasts are essential.

All in it together

The 'butterfly effect' – a metaphor used to explain one aspect of chaos theory – neatly demonstrates just how tricky a meteorologist's job is. According to the butterfly effect, something as delicate and apparently inconsequential as the flapping of a butterfly's wings in one part of the world can set off a chain reaction that changes weather conditions far away in another continent. In other words, the whole swirling atmospheric mass is so sensitive to infinitesimal changes within it that everything affects everything else.

Pity the poor old meteorologist, but this does explain why weather forecasting is not an exact science. Air temperature, air pressure, atmospheric humidity, rainfall and wind all interact with each other to determine what the weather will be. Using all the armoury of measuring, recording and analytical devices at their command, meteorologists can give fairly accurate forecasts for five or six days ahead. For longer periods, the best we can currently hope for is general trends.

INNOVATOR

British scientist Robert Hooke (1635–1703) was ahead of his time. As well as inventing and improving on various weather-recording instruments, including rain gauges, the wheel barometer and devices for measuring humidity and wind speed, he also became convinced that such phenomena as fog, storms and hurricanes were caused by changes in air pressure – and thus laid some of the groundwork for modern meteorology.

Sifting and analysing

The way in which meteorologists handle data they receive can be simple – or highly sophisticated:

- The most straightforward approach is the (rather obvious) 'persistence' method. This is based on the idea that the weather today will be much the same as the weather yesterday. This may sound simplistic but, in regions with fairly consistent weather patterns, it's a fairly reliable method – and thus perhaps hardly counts as 'forecasting' at all. In countries where the weather changes from day to day, or even hour to hour, it would not be effective.
- Getting a bit more analytical, the 'trends' method involves observing and recording particular weather features, and doing numerous mathematical calculations based on the data

collected to predict how the weather will change over a period of time. For example, if a warm front is travelling at a consistent speed, meteorologists can predict where that warm front will be in a given period of time. This focused method works well for short-term forecasting.

- The most sophisticated option is computer analysis, involving highly specialized supercomputers – some of which can get through 200 quadrillion calculations per second (that's 200 with 15 zeros). Vast amounts of meteorological data are collected and fed into these computers, which then solve a series of mathematical equations to come up with a weather map. Excellent though it is, this form of forecasting is only as good as the information received. If there's a gap in data – say, because the information is inadequate for more inaccessible areas like mountains or oceans – prediction will be less accurate.

Compare and contrast

In weather forecasting, once is never enough. When the supercomputers do their mathematical calculations, they produce what's called a deterministic forecast – a single picture that does not allow for minute changes that could affect the outcome... like butterflies flapping their wings on the other side of the world.

For greater accuracy, meteorologists produce what's known as an 'ensemble' forecast. The supercomputer reruns the deterministic forecast several times, using a slightly different starting point each time. This results in a number of different predictions. If all the results are fairly similar, the ensemble forecast is reasonably accurate. If, on the other hand, a range of very different results emerge, the forecaster will know that the prediction is less reliable.

RELIABLE SOURCE?

'It is best to read the weather forecast before praying for rain.'

Mark Twain (1835–1910)

Gizmos and gadgets

Here are some of the main devices – some hi-tech and some more homespun – that are used to measure and record weather conditions:

Aircraft With advances in technology, aeroplanes with large crews have been able to fly high in the sky and take direct recordings of weather conditions at altitude. Although they have largely been superseded by satellites, aeroplanes are still used for hurricane analysis – with their brave crews flying into the eye of the storm.

Anemomete Used to assess wind speed and direction, the most common type is the cup anemometer, consisting of three hollow cups on a vertical shaft. Different wind speeds cause the device to rotate at different rates, and these are then measured and recorded.

Balloon Weather balloons have been around since the late nineteenth century. The only problem was getting the information out of them – forecasters had to wait until a balloon floated back to land of its own accord, and this could be miles from its original launchpad. Modern weather balloons, 'radiosondes', have radio transmitters that send a constant stream of information back to Earth, such as the temperature of the air at different altitudes. They can also exchange information with weather satellites, which increases their usefulness. Floating to heights of up to 30,480 metres (approx. 100,000 feet), balloons can go where other recording devices cannot – over otherwise inaccessible oceans, for example.

Radar Developed for military purposes during World War II, radar (Radio Detection and Ranging) works by bouncing radio waves off objects and detecting the returning echoes. Its usefulness as a meteorological device was soon discovered, and now it's used to track storms, hurricanes and tornadoes, and to analyse clouds and predict rainfall. Damp air in clouds reflects the radio waves back to the radar; the damper the air, the stronger the return signal, and this allows forecasters to calculate where it's likely to rain.

Rain gauge This does just what it says – it gauges rainfall, either by measuring the weight or the volume of any precipitation. There are several different types and, although their simple construction should make them foolproof, things can go wrong – for example, snow or ice can block the funnel so that readings become inaccurate.

Satellite The real hi-tech option, weather satellites offer the most advanced method for recording atmospheric conditions. There are two types: geostationary satellites, located at about 35,680 kilometres (approx. 22,200 miles) above the surface, that travel at the same speed as the Earth turns, so they are always above the same place; and polar orbiting satellites, circling from pole to pole and travelling at a lower altitude so that they give a more detailed picture of weather conditions closer to the Earth. Satellites provide information on temperature, moisture levels, ozone distribution and solar radiation.

Sling psychrometer Used to measure relative humidity, this consists of two thermometers. One measures the dry air temperature, while the other is wrapped in a damp fabric wick. When the psychrometer is spun around, the evaporation from the wet wick causes the temperature to drop. The difference in the readings between the two thermometers indicates the level of humidity in the air.

Stevenson Screen A box-like shelter that looks a bit like a beehive and houses meteorological instruments for taking readings such as surface temperature. To avoid the strong rise and fall in temperature at ground level, the screen is raised up on legs. It has louvred sides to allow air to pass through and is painted white to reflect heat because it's the temperature in the shade, not sunshine, that is measured. The Stevenson Screen was invented in 1864 by the engineer and meteorologist Thomas Stevenson (1818–87), father of the famous author Robert Louis Stevenson.

Wind sock Not exactly hi-tech, this conical tube of fabric, open at both ends and mounted on the end of a long pole, helps airline pilots to gauge wind direction and strength. When the wind blows, it flows through the sock, lifting it so that it points away from the wind and thus shows the direction from which the wind is coming.

High standards

Air pressure and temperature are crucial ingredients in the make-up of the weather, but measuring these is not that simple. Pressure and temperature vary with altitude, so meteorologists have had to agree on a common standard:

- Air pressure is measured at sea level and is recorded in units of pressure called millibars; the average air pressure at sea level is 1,013.25 millibars.
- Air temperature is measured at a height of 1.5 metres (5 feet) above ground. So if the weather forecaster says the temperature will be 27°C (80°F) tomorrow, that refers to the temperature of the air swilling about at roughly that height.

HOT AND COLD

It was all very well having thermometers to measure temperature, but to give meaning to the temperature scale, it had to be linked to temperatures in the natural world. The German physicist and engineer Daniel Gabriel Fahrenheit (1686–1736) took the approximate temperature of blood as the top end of his scale at 100°F – his northern European compatriots did not expect air temperatures to rise any higher than this. The temperature at which fresh water froze then became 32°F. The Swedish astronomer Anders Celsius (1701–44), on the other hand, took the freezing point of water as his 0°C, and the boiling point as 100°C.

SIGNS AND SYMBOLS

Weather charts have a language all their own – and it's fun to learn how to read them. Wind is measured in knots and, on a weather chart, winds are represented by lines called 'barbs' that look rather like little arrows with feathers at the tail. The barbs point in the direction in which the wind is blowing and the 'feathers' – or short prongs protruding from the side of a barb – indicate its speed. If the wind is really strong, a pennant-shaped triangle is added to the side of the barb. Here's the maths:

1 knot = 1.9kph (1.15mph)
Full-length prong = 10 knots
Half-length prong = 5 knots
Pennant = 50 knots

Prongs and pennants can be added up to give an actual wind speed. So, for example, a full-length prong and a half-length prong show a wind speed of 15 knots.

3. The Armchair Meteorologist

∽

ARMED WITH GREATER understanding of the weather and its causes, you can now do a bit of your own forecasting. The choice is yours – stick to careful observation backed by your new-found knowledge; let your inner child have some fun doing a few experiments and making some simple instruments; or go the whole hog and set up your own mini-weather station.

Judging the temperature

To measure air temperature, you will need a simple gadget – an air thermometer (or perhaps two, for more accurate comparison). These are available online, or you can buy them from specialist shops and some science museums. Now get creative… you may be surprised by some of the results:

- Measure the temperature at different heights, say 5 centimetres (2 inches) above the ground and 2 metres (approx. 6½ feet). What's the difference, and how does it change in different weather conditions and at different times of the day?
- Take temperature readings close to buildings and out in the open – what's the difference?
- Find the maximum and minimum temperature during each 24-hour stretch. Take two readings – one at 2pm for the maximum temperature, and one as early as possible in the

THE ARMCHAIR METEOROLOGIST

morning, when temperatures are similar to the night-time minimum. Record the temperatures over a number of days and look for similarities, differences and patterns.

The way the wind blows

When someone observes that it's windy, you can smugly remind yourself that you know better – that isn't just wind, it's a mass of high-pressure air rushing in to fill an area of low pressure. A windsock – of the kind you see at airports – and a wind vane (see illustration on page 14) are two simple devices you yourself can construct to detect the direction of the wind. For the wind vane, ensure that the compass points on the two fixed cross-bars are accurate – use a compass to help you get it right. For a windsock see page 49 for details. Mount your sock or your vane high up, where they can catch the wind unimpeded.

Mirror, mirror…

… in this case, not on the wall but the ground. This satisfyingly simple method will help you assess how cloudy the sky really is. To start with, you'll need a large mirror.

- Using a ruler and a pen, chinagraph pencil or wax crayon, mark out sixteen equal squares on the glass surface (make sure it's not a family heirloom before doing so!).
- Lay the mirror flat on the ground, somewhere where you can see the whole sky reflected.
- Count the number of squares with cloud in them, then divide this number by two to convert your total to 'oktas'. Oktas are the units of measurement used to record the amount of cloud

cover; each represents one-eighth of the sky. On a sliding scale of 0–8, 0 oktas at one extreme signifies a totally clear sky, while 8 oktas means it's completely overcast, with the oktas in between giving you varying amounts of cloudiness.

Testing the water

To record rainfall, any container in which you can collect water and measure its depth will do. One problem, though, is that some of the rainwater may have evaporated before you get to do your measuring. For more accurate results, buy a purpose-made device – or mark on measurements at the side of your container before placing the container outside. Situate your gauge in an open space, well away from trees or buildings.

- Check your gauge and measure the amount of rainfall every day at the same time.
- Arrange a number of jam jars or other simple rain collectors in different spots – in the open as well as under trees and near buildings. How much of a difference do trees and buildings make to the amount of rain that reaches the ground?

Waterworks

- Although snow isn't frozen rain – snowflakes form in clouds – it is made from the same basic substance as rain: water. Snowfall of about 25 centimetres (10 inches) roughly equates to 2.5 centimetres (1 inch) of rain. You can check this out by collecting snow in a measuring jug, noting its depth, then melting it to see how much water it makes.
- Raindrops in a shower are bigger than those in drizzle. Compare their sizes by letting them fall onto coloured blotting paper, then quickly measure how big they are.

- It's easy enough to measure the depth of snowfall – just find a stretch of level snow, avoiding any patches where it has piled up unevenly, and insert a stick or short length of wood into it. Make a mark on the stick in line with the top of the snow, then compare this measurement with a ruler to get the depth in centimetres or inches.

A pressing matter

For this experiment to see how air pressure changes with altitude, you'll need a barometer and a convenient hill – the higher the better. Or, if you're surrounded by urban concrete, a skyscraper or tall church tower will do instead. Take barometer readings at different heights, at ground level and then at various levels on the way up. Keep a note of your readings, and the time of day at which you took them, for later comparison. To build your own makeshift barometer, see page 34.

Home weather station

It's perfectly possible to set up your own small-scale weather station to track local weather, but it will involve some expense. The basic beginner's kit you'll need is:

- a *barometer* for reading air pressure.
- a *thermometer* for reading maximum and minimum temperatures and everything in-between.
- a wet- and dry-bulb *hygrometer* for reading humidity (see diagram).

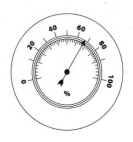

You'll also need a screen to house your instruments – your DIY Stevenson Screen. You could use a wooden box with no base, or you could buy a louvred screen. Paint it white, to reflect heat radiation, and keep it clean. Set it up in the open, away from trees and buildings, and make sure it is secure and level. Place your instruments inside it and take regular daily readings, at the same time each day or, ideally, twice a day.

Where to get your kit

Meteorological organizations and museums may offer some equipment, or look online. Organizations and websites are also a good source of inspiration for experiments and homemade equipment. You could start with the following:

www.metoffice.gov.uk
www.weather.gov
www.rmets.org/weather

Put it down on paper

Don't waste all that research – log your data in a special note-book, and compare your notes to look for patterns and to gain new understanding of your local weather. Throughout this book there are 'Do It Yourself' hints, to show you how you can keep track of the weather without having to turn your home into the BBC weather station.

SECTION 2:
HOW WEATHER WORKS

4. The Structure of the Atmosphere

෴

WE CAN'T TOUCH it, we can't smell it, but it's all around us and it's vital to our survival: it's the Earth's atmosphere. The atmosphere is our planet's comfort blanket, regulating its temperature and protecting it from radiation. When we gaze up into its blue vastness, it may not seem as if much is going on there, but don't be fooled – it's a busy place where the constant but invisible interaction of natural forces creates our changing weather.

Gasbag

One clue to the nature of the atmosphere may be found in the word itself. Head for your nearest dictionary and you will see that it is derived from the Greek words *atmos* (vapour) and *sphaira* (sphere). In other words it's just a great big bag of gas. In fact, the atmosphere is made up of a number of gases that together constitute what we call 'air'.

The air we breathe

The oldest gases in the atmosphere include carbon dioxide, nitrogen, methane, and the so-called 'noble gases' such as helium and argon. Water vapour has been there a long time too. But it wasn't until the appearance of primitive bacteria, millions of years ago, that the vital component of the air we breathe first materialized.

Using the power of the sun in a chemical process called photosynthesis, these bacteria absorbed carbon dioxide from the atmosphere and gave off a new gas: oxygen. Some of this oxygen was converted to ozone, which then formed a protective shield against the sun's harmful ultraviolet light, making it possible for more complex life forms – including human beings – to develop.

Now, around 78 per cent of the air in the atmosphere is made up of nitrogen and 21 per cent of oxygen. The remainder is composed of other gases, water vapour, aerosols (solid particles such as dust, viruses and bacteria), and minute quantities of highly reactive molecules such as the hydroxyl radical (OH). OH is important because it helps to clear the atmosphere of pollutants such as hydrocarbons.

Pulling power

- The atmosphere is held in place by the Earth's gravitational pull.
- The atmosphere stretches up to around 10,000 kilometres (6,214 miles) above the planet's surface, but even so, it is still smaller than the Earth's diameter, which is almost 13,000 kilometres (approx. 8,000 miles) wide.

- Because the pull of gravity is strongest closest to the Earth, most of our air is found approximately in the lowest 17 kilometres (10½ miles) of the atmosphere.
- The further out into the atmosphere you go, the thinner the air becomes because of the weakening power of gravity. This is why it gets harder to breathe at high altitudes, and explains why mountaineers sometimes carry oxygen tanks.

How much can you take?

On the ground, we aren't aware of the weight of the air pressing down on us because our bodies are strong enough, and contain sufficient air, to exert a corresponding pressure. When we go up in an aeroplane, on the other hand, the external pressure drops considerably. To compensate and reduce stress on the plane's fuselage, the pressure inside the cabin is lowered. When we come in to land, it is increased again to match the increasing pressure outside, which is why your ears pop and you may find yourself temporarily deaf on landing – your body has been adjusting to a change in pressure.

Don't push me!

The pressure exerted by air is traditionally measured in pounds per square inches, or psi. At 3,000 metres (10,000 feet) above the surface, air pressure is about 10lb per square inch – that's 10 psi or 4.5 kilograms for every 6 square centimetres. At sea level, this figure rises to 14.7 psi.

DO IT YOURSELF!

To test air pressure you need a barometer, but you can make your own basic barometer with just the following items:

- A balloon
- Glass or jar (wide mouth/ opening)
- Drinking straw
- Rubber (elastic) band
- Tape
- Piece of paper
- Scissors & pencil

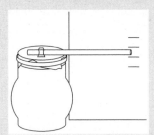

Inflate the balloon to stretch it, then allow it to deflate. Then chop off the balloon at its middle and discard the end with the neck that you blow into. Stretch the half that you are left with over the top of your lidless jar so that it covers the top and seal with an elastic band. By doing this you're capturing air at a certain pressure.

Now tape a straw onto the balloon 'lid' approximately a quarter of a way along the straw's length (see diagram). Place the jar by a wall with a piece of paper stuck to it, and mark where the initial position of the straw is so you can monitor its movement. The balloon indicates changes in air pressure around the jar – higher air pressure pushes the balloon into the jar and makes the straw rise; low air pressure makes air inside the jar expand and the balloon lid rise, moving the straw down.

By marking where the straw moves to at regular intervals during each day over the course of a week, you will start to see patterns – with sunny days the straw should move up with the higher air pressure, and with rainy days the straw will drop lower. So, with this handy homemade gadget you can predict a change in the weather!

UNDER THE WEATHER

If you're finding it hard to focus and have forgotten where you put your specs, why not blame it on the weather? Atmospheric pressure fluctuates all the time, and research in the Ukraine suggests that even slight changes can affect our ability to concentrate and our short-term memory.

All systems go

As well as being the place where weather happens, the atmosphere is like a well-regulated house where temperature and air quality are controlled by central heating, air conditioning, humidifiers and good loft insulation. The atmosphere...

- provides oxygen-rich air for us to breathe.
- recycles the water vapour in the air, condensing it into life-giving rain.
- allows in enough of the sun's heat and light to sustain life, while at the same time protecting the Earth from harmful ultraviolet radiation.
- insulates the planet from the freezing temperatures of outer space.

Give me five

The Earth's atmosphere may be divided into five layers or zones, as follows:

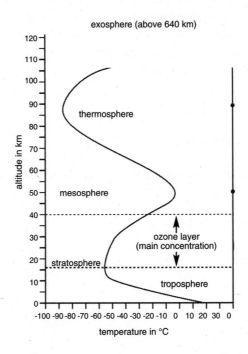

exosphere (above 640 km)

1. The troposphere

This first layer, closest to the Earth, stretches up to 8–17 kilometres (5–11 miles) above the surface and is where changing atmospheric conditions control our weather. It contains the highest density of air, holding about half of the planet's atmospheric gases. Acting like a giant storage heater, the Earth's surface retains heat from the sun and reflects this out into the troposphere, so that the air in this zone is warmest closest to the surface.

2. The stratosphere

Lying 17–50 kilometres (11–31 miles) above the Earth, this zone contains the ozone layer, at an altitude of about 25 kilometres (15½ miles). A reactive type of oxygen, ozone absorbs many of the sun's harmful rays, thus protecting the planet. As it absorbs these rays, the ozone heats up so that in the stratosphere – unlike the troposphere – the temperature increases with altitude. There are few clouds in this sphere.

3. The mesosphere

Reaching up to 85 kilometres (53 miles) above the Earth, the mesosphere – or 'middle' sphere – contains no ozone and is decidedly chilly – it's the coldest layer of the atmosphere.

4. The thermosphere

The outer edge of the thermosphere – the 'hot' sphere – lies up to 1,000 kilometres (621 miles) above the Earth. The air here is extremely thin. Unlike the cold mesosphere, temperatures here begin to heat up again because – with no ozone to soak it up – solar radiation is more powerful here, leading to highs of more than 1,700°C (3,092°F).

5. The exosphere

This 'outer' sphere is where the Earth's atmosphere meets outer space. The exosphere extends above the thermosphere, and some scientists think it stretches up to 10,000 kilometres (6,200 miles) above the planet.

5. Blowing Hot and Cold: Heating and Cooling Systems

ॐ

YOU MAY NOT be able to see everything that's going on in the atmosphere, but there's never a dull moment up there. The invisible air masses within it are constantly on the move – and it's this continual movement and change in the air that explains much of the weather. Here's (a bit of) the science behind it all.

Up a bit, down a bit, left a bit…

Our atmosphere is like a vast heating and cooling system, and it's the heat of the sun that powers it. Here – in the simplest terms – is how it works:

- The sun warms the Earth's surface and, in turn, the surface – like some giant electric storage heater or hotplate – absorbs this heat, then radiates it out to warm the air above.
- The warm air radiating from the Earth expands and rises. As it does so, the air pressure decreases. So an area of low pressure is an air mass that is on the up. But as it rises, the air cools and contracts, and the vapour within it condenses into clouds. So if the weather forecaster mentions 'low pressure' be prepared for grey skies and rain.

- Now the reverse process begins to happen. The cool, compressed air starts to sink back down again and the pressure mounts: so an area of high pressure is an air mass on the way down. As it sinks, the air gets warmed by the Earth again, starts to expand once more, and retains more water. So if the forecast includes the words 'high pressure' you should be in luck because the weather is likely to be fine, clear and dry.

Exchange rate

If you stood on a ladder to decorate a ceiling in a heated room, it's likely that it would be warmer up there than on the floor because – as we all know – hot air rises. To stop that hot air rising up and right out of the roof, it's a good idea to have loft insulation. The atmosphere has some insulation in its lowest layers – greenhouse gases and clouds – but as it goes higher than this the air loses heat. Dry, rising air cools and sinking air warms up at a constant rate of about 10°C (50°F) per kilometre (0.6 miles) – which explains why air at the top of a high mountain is pretty chilly. It's more complicated with moist air, though, because condensation alters the cooling rate.

Night and day

Heated by the sun during the day, the ground warms the surrounding air. By night, however, as the Earth rotates away from the sun, what warmth the surface has absorbed by day escapes into the atmosphere, especially if the night is clear with no nice, cosy cloud cover to act as insulation. This helps to explain the difference in temperature between night and day.

Fasten your seatbelts...

... it's going to be a bumpy ride. You may not be able to see it, but bumpy rides give you a clue to small changes in air pressure. Even if you were still living in the Dark Ages and thought that the Earth was flat, you would have to agree that there are hills and mountains on its surface. When the sun's rays hit this ridged, furrowed surface, they won't reach all parts to the same degree – for example, the sunny side of a hill gets more heat than the shadowy side. And that means that the surrounding air is also heated unevenly, resulting in pockets of varying air pressure – and 'bumpy air' for air passengers flying low over high hills or mountain ranges.

Cruise control

Thermals – spiralling columns of warm, rising air – are also caused by the uneven heat given off by uneven terrain. Birds of prey, such as hawks and eagles, take advantage of these thermals to give themselves lift, so they can spread their wings and glide and not waste energy flapping. Glider pilots make use of the same phenomenon. Because thermals are generated by the heating effects of the sun, they do not occur at all times of the day or year. This explains why gliding birds usually only take to the skies in sunny weather, and why they generally become active only after ten o'clock in the morning, when the thermals have become strong enough to give them the necessary lift.

Land and sea

Land surfaces respond more quickly to the sun's heat and absorb and radiate it more readily than the surface of the oceans, which remain at a more constant temperature. This, along with the fact that the surface of the land is more irregular, is one reason why thermals are rare over seas (and why air passengers may get a smoother ride at low altitude flying over the water).

ANCIENT WEATHERMEN

Aristotle, the Greek philosopher, had his own ideas about the causes of weather – and he wasn't that far off the mark. According to him, earth and water, when warmed by the sun's heat, combined to produce 'vapours' that were the cause of rain and snow, and 'exhalations' that created the wind. Leonardo da Vinci held to the theory that there were two types of air: 'fire air' that fuelled flames and sustained life; and 'foul air' that had the opposite effect. When oxygen was discovered, it was labelled 'fire air'.

WATCH YOUR LANGUAGE!

Impress your friends and astonish your dinner guests by flashing your knowledge of meteorological terms:

Anticyclone The zone of highest pressure within a given area. On weather maps it is marked with an H.

Convection The process that occurs when air is heated from below, and rises.

41

Convergence A process that occurs when air masses flow into the same area from different directions. With so much air flowing into the one area, the only way out is up, so the incoming air pushes up the layer of air above.

Cyclone The zone of lowest pressure within a given area. On weather maps it is marked with an L.

Dewpoint The temperature at which cooling air becomes saturated and the water vapour it contains condenses into tiny droplets of water. In the sky, this means clouds.

Depression This could be what everyone starts to feel if you hold forth on the subject of weather for too long – but it's also another name for a low pressure area, or a cyclone.

Isobar A line on a weather map joining up points of the same pressure (see diagram).

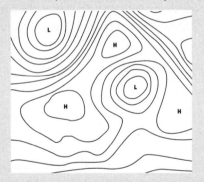

Precipitation A swanky name for water falling from the sky – in other words, rain, hail or snow.

Thermal A rising, swirling column of warm air. Check out the presence of thermals by seeing what gliding birds are up to, or watch out for cumulus clouds (the kind that look like big fluffy balls of cotton wool) as these form at the top of thermals.

SECTION 3: AIR AND WINDS

6. Putting on a Brave Front: Weather Fronts Explained

∽

COLD FRONTS, WARM fronts – we've all heard the terms, but just what do they mean? Knowing where fronts lie and what causes them is an essential piece of information for any forecaster (amateur or professional) who wants to predict the weather.

Round trip

Warm air rises, cold air sinks, causing different levels of air pressure. All clear so far. But it doesn't end there. Air masses not only move vertically, but they move horizontally too. It's all part of the atmosphere's thermostatic control system that seeks to even out air temperatures around the globe. Air at the equator – the hottest part of the Earth – is heated, causing it to rise and move out towards the poles – the coldest places on Earth – cooling as it does so, and eventually sinking. This cool air is then forced to flow back to the equator to replace the hot air that is rising there. So – give or take the effect of the Earth's spin and various other factors – areas of circulating air are generated. As in life, so with the weather – what goes around comes around.

43

Wet or dry?

The surface over which air travels can affect it too: 'maritime' air that travels over oceans and seas will pick up extra moisture, while 'continental' air that moves over land will remain relatively dry. Flash your new knowledge by tossing the following terms casually into a conversation about weather:

Tropical maritime	A warm, moist air mass
Tropical continental	A warm, dry air mass
Polar maritime	A cold, fairly moist air mass
Polar continental	A cold, dry air mass

Don't be affronted

Cold, polar air can't mix with warm, tropical air, so the transitional area between the two masses is called a 'front':

- A cold front heads an advancing mass of colder air that is going to replace warm air. On a weather map, it's marked by a line of blue triangles showing the direction in which the front is travelling.

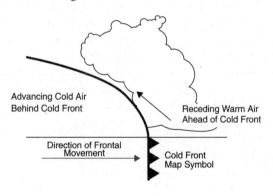

- A warm front heads an advancing mass of warmer air that will replace cold air. On a weather map, it's shown as a line of red semi-circles indicating its direction.

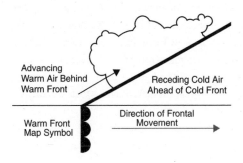

Advancing Warm Air Behind Warm Front

Receding Cold Air Ahead of Cold Front

Direction of Frontal Movement

Warm Front Map Symbol

A change in the weather

Fronts are important to forecasters because they signal a change in the weather. Both types usually bring increased humidity, with clouds and rain, but while a cold front is followed by a drop in temperature, a warm front means warmer weather.

DON'T RAIN ON MY PARADE

If one front follows another in fairly rapid succession, this can lead to changeable weather – as often happens in winter in Britain, for example, which is notorious for its unpredictable climate. But British summer weather can be equally capricious. After extensive consultation with meteorologists, 2 June 1953 was chosen as the most reliably sunny day for the crowning of Queen Elizabeth II, an occasion that would include the young queen riding in the golden state coach past cheering crowds, many of whom had spent the previous night sleeping on the pavements to stake their places. Naturally, British weather being what it is, when the great day came, it poured with rain.

How's the weather?

It isn't just the British who suffer such changeable weather, however. American writer Mark Twain is said to have proclaimed: 'If you don't like the weather in Buffalo, just wait five minutes'. Many other localities have been similarly named and shamed – from Chicago, Boston and Washington in America to Melbourne in Australia – with waiting time ranging from five to twenty minutes to a whole day.

7. Getting the Wind Up: All About Winds

❧

BREEZES, GUSTS, GALES – call them what you will, winds are just bodies of fast-moving air. But what gets them going in the first place, and why do they blow at different speeds?

Filling the vacuum

Winds are a response to differences in air pressure. They blow from high-pressure areas, where cooling air is sinking, to low-pressure areas, where warming air is rising, in an attempt to even out the pressure. Given the choice, winds would just get on with their job and flow directly from one area to the other. But the friction caused by the surface over which they travel, the rotation of the Earth and something called the Coriolis Force all interfere and deflect the wind's direction.

As a result, in the Northern Hemisphere, a wind will swirl around, down and out in a clockwise – or *anticyclonic* – direction from a high-pressure area; and around, up and into a low-pressure area in an anticlockwise – or *cyclonic* – direction. South of the equator it's the other way round. Generally, the greater the difference in air pressure, the stronger the wind will be. Easy, isn't it?

47

'When the wind is in the east
It's good for neither man nor beast.
When the wind is in the north
The skilful fisher goes not forth.
When the wind is in the south
It blows the bait in the fish's mouth.
When the wind is in the west
The weather is at the best.'

Traditional rhyme

Spinning top

Think of planet Earth like a spinning top, delicately balanced at a slight angle. Air movement between the hot tropics and the cold poles would normally flow directly from the centre to the north and south respectively, and back. But because of the planet's spin, the natural flow is twisted off course to travel diagonally rather than directly up and down.

Read the signs

Winds generally follow the isobars (lines joining points of the same pressure). When the isobars on a weather chart are close together, it means that there is a steep change in pressure – so you can expect a blustery day. Close isobars plus low pressure mean it's going to be wet too.

Record wind

On 10 April 1996 the fastest wind speed recorded on Earth was measured by an automated weather station on Barrow Island, Australia, during Tropical Cyclone Olivia. The gust reached 408kph (253mph).

DO IT YOURSELF!

It's easy to test the direction of the wind using a windsock and a compass. To make one yourself gather up the following items:

- Any plastic bottle
- Cloth ribbons
- Sellotape
- Cutting tool
- String
- Hole punch

Cut the top and bottom of the bottle, which will leave a ring of plastic – make sure it's a few centimetres thick. Then you'll need to punch two holes at the top of your plastic ring at opposite ends and put string through both holes so that you can tie up the windsock. Punch four more holes at the bottom of the ring at equal distances and tie some ribbons in these holes. The windsock will then need hanging up, probably on a pole so that the wind can get to it from all directions. When the wind starts to blow the sock and ribbons will lift, and if you check your compass you'll see which direction it's coming from.

Ship ahoy!

Long before any talk of alternative green power, sailors were already using a source of natural energy to propel them across the seas: wind. However, the wind was an unpredictable, uncontrollable force and sailing ships were at its mercy. An accurate method of assessing its speed was necessary, but sophisticated, scientific equipment was not yet available.

In 1805, Rear-Admiral Sir Francis Beaufort of the British Royal Navy came up with a bright idea. By observing how frigate sails reacted in different wind speeds, he invented a scale for measuring the force of the wind – a wind might be so light, for example, that it gave only enough power for the ship to be steered, or so strong that it would shred the sails. Known as the Beaufort Scale, it was adopted by the Royal Navy in 1838.

When steamships appeared, the behaviour of the sea was also taken into account and, later, conditions on land. The Scale originally had thirteen categories from 0 (calm) to 12 (hurricane force). In 1955, the United States Weather Bureau added a further five, forces 13 to 17, to grade hurricane-force winds. The Beaufort Scale is still used today, alongside more scientific methods.

Here are the original categories of The Beaufort Scale, expressed in knots and miles (for landlubbers, a knot is a unit by which a ship's speed is measured, and is equivalent to about 1.8kph (1mph).

Force	Average speed knots/mph	Description	Effects
0	0/0	Calm	Calm, glassy sea; smoke
1	2/2	Light air	Wind direction shown by smokedrift but not wind vanes
2	5/6	Light breeze	Ripples on sea; wind felt on face, leaves rustle, vanes moved by wind
3	9/10	Gentle breeze	Smooth wavelets; small twigs in constant motion, light flags extended
4	13/15	Moderate breeze	Numerous whitecaps on sea; dust, leaves and loose paper raised, small branches move
5	19/21	Fresh breeze	Many whitecaps on sea, some spray; small trees sway
6	24/28	Strong breeze	Larger waves, whitecaps everywhere, more spray; large branches move, whistling in phone wires, umbrellas difficult to use
7	30/35	Near gale	White foam from breaking waves blown in streaks; whole trees in motion; walking inconvenient
8	37/43	Gale	Wave crests begin to break into spindrift; twigs break off trees, hard to walk

Force	Average speed knots/mph	Description	Effects
9	44/51	Severe gale	High waves, heavy swell, spray may reduce visibility; chimney pots and slates removed, large branches break
10	52/60	Storm	Very high waves with overhanging crests, sea white with blowing foam; trees uprooted, structural damage
11	60/69	Violent storm	Exceptionally high waves; widespread damage
12	64+/74+	Hurricane	Sea white, air filled with foam, visibility greatly reduced; widespread damage, rarely experienced on land

WATCH YOUR LANGUAGE!

Gust A wind that lasts from a few seconds to a few minutes.

Which way does the wind blow?

When the weather forecaster talks about a 'northerly' or 'westerly', this refers to the direction the wind is coming *from* – not where it is heading.

8. Winds of the World

～

HURRICANES ARE GIVEN their own individual names, ranging from Arthur to Joaquin to Wendy. While less powerful winds don't quite have the same status, some localized ones are so regular and so predictable that people felt they deserved more than just the description 'breeze' or 'gust' or 'gale.' Here are a few winds with names, some familiar, some less so...

Monsoon

In most people's minds the monsoon conjures images of India and Southeast Asia, along with copious quantities of rain. Technically, however, a 'monsoon' is a wind system that dominates the climate of a wide region, is found in various parts of the world and comes in both 'wet' and 'dry' forms. The Asian monsoon is simply the most famous example.

In Asian summers, the land heats up more than the surrounding ocean (ocean temperature being more stable than that of land), creating a large low-pressure area of warm, rising air over north-central Asia and a smaller one over India. This activates a strong wind, as air rushes to fill these low-pressure areas.

Blowing in from the Indian and Pacific Oceans, it carries moisture with it. Warmed by the hot land and forced upwards by mountain ranges such as the Himalayas, the mass of fast-moving air rises, cools and condenses, shedding its water load in the form of torrential rain, in the wet monsoon season.

During the winter, however, the airflow goes the other way. The land cools rapidly, the air sinks and a large high-pressure area known as the Siberian High forms over north-central Asia, with a smaller one over India. The drier, colder wind is now drawn out to sea, in the dry monsoon season.

Trade winds

A different airflow from the one that creates the monsoon is responsible for the 'trade winds'. Around the equator and in the tropics, warmed air is constantly rising high into the sky and flowing out towards the poles, while a steady stream of winds flows in towards the equator. Due to the effect of the Earth's rotation, they blow from the northeast in the Northern Hemisphere and the southeast in the Southern Hemisphere, converging on the equator. Because these winds were the ones that propelled trading ships from Europe to the Americas, they became known as the 'trade winds'.

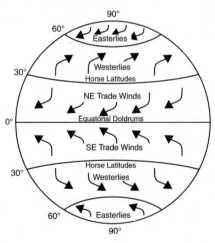

In the doldrums

In the band between 5 degrees north and 5 degrees south of the equator, the air is still and there is little wind. In the days of sail, it was bad news for a ship to get stranded here and despondent sailors called the region the doldrums, after an old English word meaning 'dull'.

Horse latitudes

In the so-called 'horse latitudes', between about 30 degrees and 35 degrees north and south of the equator, winds are weak. Becalmed here and fearful of running out of drinking water, the crew of sailing ships are said to have thrown overboard any horses and cattle that were dying of thirst – hence the name. These latitudes are also known as the Subtropical High, or, north of the equator, the Calms of Cancer, and to the south, the Calms of Capricorn. The irony of being without drinking water while surrounded by water – undrinkable because it was so salty – was not lost on Coleridge's Ancient Mariner:

> Water, water, every where,
> And all the boards did shrink;
> Water, water, every where,
> Nor any drop to drink.
>
> From 'The Rime of the Ancient Mariner' by
> Samuel Taylor Coleridge (1772–1834)

Katabatic winds

When a mass of cold, dense air, pulled by the force of gravity, slides down the side of a mountain to the valley below, a katabatic (from the Greek words *kata*, meaning 'downwards') wind arises. The most extreme example is found in the Antarctic, where katabatic winds can gust across the ice caps at speeds of 193kph (120mph).

Getting personal

'They called the wind Maria' – so sang one of the characters in the 1950s Broadway musical *Paint Your Wagon*. Here are some other winds with names:

Bora A northeast, katabatic wind that blows across the northern Adriatic, reaching speeds of more than 100 miles an hour.

Canterbury Northwester A northwesterly wind that sweeps down the mountains of the New Zealand Alps and onto the Canterbury plains of South Island, New Zealand.

Föhn A warm, dry wind that roars down the slopes of the Swiss Alps and that has been blamed for all kinds of nasty symptoms, from headaches to depression and suicidal urges. 'Föhn' (or 'foehn') has also become the generic name for any wind that heats up as it blows down a slope.

Habob From the Arabic word *haab*, 'to blow', a wind that blows in northern Sudan, usually in the afternoon and evening, bringing sandstorms.

Karajol A westerly wind found along the coast of Bulgaria, typically lasting just one to three days and bringing dry weather.

Levanter Blowing from the east or northeast in summer, the levanter is mild and humid along Spain's southeast coast but can reach gale-force as it funnels through the Straits of Gibraltar.

Mistral A cold, dry, katabatic northerly that travels across the Mediterranean coast of France in winter. As it is forced through valleys in the mountainous terrain, it blows faster, sometimes reaching a velocity of 130kph (80mph) and whipping up heavy seas – much to the delight of surfers.

Puelche A warm dry wind blowing from the east, moving across the Andes from Argentina to Chile. It is named after the Puelche people who live to the east of the Andes.

Reshabar Dry and warm in summer, cold in winter, a strong, gusty northeasterly that blows across southern Kurdistan.

Simoom An extremely hot, extremely dry wind that gusts across the Sahara Desert, through Palestine, Jordan, Syria and the deserts of the Arabian Peninsula, carrying clouds of sand and dust, forming itself into whirlwinds and reshaping the dunes. Luckily it doesn't usually last more than about twenty minutes but it can still cause heatstroke. The locals know what they think of it: the name 'simoom' comes from *samma*, the Arabic word for poison.

Southeaster Arising in areas of high pressure far off in the Southern Ocean, a moisture-laden southeasterly that arrives in Cape Town, South Africa, in the summer, creating the famous 'tablecloth' – the bank of cloud that drips over the top of Table Mountain. Flowing around the mountains, it then becomes trapped in the bowl between them and the sea, careering around at high speed and purging the city's air at the same time – hence its other name, the Cape Doctor. Cleansing effect apart, it often outstays its welcome: in November 1936 it blew non-stop for fifteen days, wrecking gardens and trapping staff in the upper Cable Car station on Table Mountain for five days.

A DEVILISH AFFAIR

The tablecloth that sometimes hangs over the crest of Table Mountain is supposedly not wind-blown cloud but smoke, or so an old tale would have it. Retired from his exploits at sea, an early eighteenth-century Dutch pirate by the name of Van Hunks liked to sit puffing on his pipe on the slopes of Devil's Peak, adjacent to Table Mountain. One day a stranger appeared and challenged him to a contest to see who could out-smoke the other. The two puffed away for days. Finally, conceding defeat, the stranger revealed his identity: Van Hunks' opponent was no less than the Devil himself. The two disappeared in a puff of smoke and have not been seen since – but the smoke that they created still drifts over the top of Table Mountain in summer when the Southeaster blows.

Tramontana Bringing distinctly chilly air from the Alps and Apennines, a northerly wind that blows in the Mediterranean, especially along the west coast of Italy and in northern Corsica.

Vardar A cold wind that blows from the northwest, travelling from the Vardar valley in Greece to Macedonia. It is also known as *vardarac*.

9. It's Chilly Out There: The Wind Chill Factor

∽

E VER HEARD A weather forecaster talking about a mystifying something called the 'wind chill factor'? This has nothing to do with how low the temperature is – it refers to how cold the wind makes you *feel*; the faster the wind, the colder the effect.

Not too hot, not too cold

If the weather is calm, relatively cold outdoor temperatures can still be bearable. But a wind can make the same temperature feel a lot colder. This is where 'wind chill' comes in – it's the effect of the wind on heat loss from your body – or, to be more exact, from your skin. The skin has two ingenious mechanisms for regulating body temperature:

- to cool us down, it perspires and the evaporating moisture takes excess heat with it.
- to protect us from cold, it radiates heat to create a thin, insulating layer of warm air around us.

Wind interferes with both these processes: it disturbs the skin's insulation and it speeds up the evaporation of moisture from the skin.

Arctic temperatures

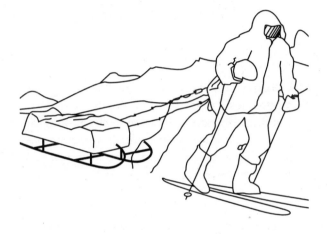

For most of us, wind chill is likely to be no more than an unpleasant inconvenience as we head hurriedly back to our cosy, centrally heated homes. But prolonged or severe exposure to a combination of wind and low temperature can be dangerous to your health. Early research into the wind chill factor was based on conditions in Antarctica, where frostbite is an ever-present danger. Researchers discovered that the speed of the wind – not the air temperature – largely determined whether or not a person would get frostbite. For example, someone could work in -40°C (-40°F) but a wind of only 5-7kph (3-4mph) could make this much more difficult.

DO IT YOURSELF!

The equation for working out the wind chill factor is rather complex, so instead you can buy an anenometer that calculcates wind chill factor for you. You'll know whether you need to wrap up warm before heading out!

In cold blood

Low temperatures, aggravated by wind chill, can lead to such hazards as hypothermia and frostbite. And if you are of a certain build or of a certain age, or have certain health conditions, you will be more susceptible to the cold anyway so need to take extra care.

- If your forehead is exposed to a wind chill of less than -50°C (-58°F), you could be unconscious in minutes.
- If your skin is exposed to a wind chill of -75°C (-103°F), it may freeze within thirty seconds.
- Children and older people are at greater risk from cold because their bodies are less efficient at regulating temperature. Some Canadian schools even have a policy of keeping children indoors when the wind chill reaches a certain level.
- If you are tall and thin, you have a greater area of skin in relation to your mass than shorter, broader types, so there's more skin to be exposed – hence you're likely to feel the cold more readily.

- Poor circulation, damaged blood vessels, smoking, drinking alcohol and taking certain types of medication interfere with the way blood, and therefore heat, is transferred around the body, so affected individuals may be more susceptible to the cold.

10. Up, Up and Away:
Jet Streams

〜

W AY UP IN the atmosphere, currents of fast-flowing air zoom around planet Earth. Thousands of kilometres long, hundreds of kilometres wide and just a few kilometres thick, they steer our weather systems and, if you're flying, can give the aircraft you are in the extra push it needs to get you home early. These mysterious aerial forces are known as 'jet streams'.

Get a move on!

Jet streams occur about 10–12 kilometres (33,000–39,000 feet) above the surface of the Earth, where masses of cold polar air and warm tropical air meet. The different densities of these

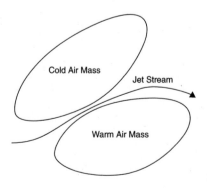

masses – cold and heavy, warm and light – mean a difference in pressure, and this forces air to rush at high speed from one to the other, creating the fast-moving bands of wind known as the polar and subtropical jet streams.

Instead of flowing directly from warm to cold, however, jet streams are deflected by the Earth's spin and travel along the boundaries between the two masses, taking a meandering path. The main jet streams usually flow from west to east but sometimes they can change to almost north to south – all part of the atmosphere's clever mechanism to transfer heat from the equator to the poles, and cold air from the poles to the equator.

The position of the jet streams is a clue to where the strongest contrast in surface temperature lies. The contrast is greatest in winter, so this is when the streams are fastest, accelerating to 482kph (300mph) or more – over southwest Scotland, for example, speeds of 643kph (400mph) have been recorded.

WATCH YOUR LANGUAGE!

Jet streak The tearaway of the skies – a pocket of extra-speedy wind within a jet stream that can blow at more than 320kph (200mph).

The answer is blowing in the wind

Because they are strong currents of fast-moving air, jet streams have the ability to steer different weather systems around. Knowing where the streams are, and how rapidly they are blowing, helps forecasters to predict what the weather is going to do. In Britain, for example, rain-bearing depressions (centres of low pressure, also known as cyclones) move in from the Atlantic.

However, if the Polar Front Jet Stream – the one that affects British weather – changes its usual easterly course it can steer such depressions away, or even block them altogether. Gazing into their crystal balls, forecasters are even worried that, should global warming increase sufficiently, this jet stream will move northwards, producing drought conditions over much of the UK in the summer. The winds in a jet stream do not always blow at a constant speed and at the points where they accelerate or slow down areas of low or high pressure can form, further influencing our weather.

The Polar Front Jet Stream

Forming at a latitude of 60 degrees north during the winter between 9 and 13 kilometres (30–43,000 feet) above the Earth, this strong band of winds controls the weather in the high latitudes of the Northern Hemisphere.

The Equatorial Jet Stream

Located at about 7 degrees north of the equator at an altitude of up to 15 kilometres (50,000 feet), this is another strong band of winds, which move from the east, from Asia across Africa, unlike the other streams, which flow from the west. It is strongest in July and August, in the Northern Hemisphere's summer.

The Subtropical Jet Stream

Lying 30 degrees north and south of the equator at an altitude of around 13 kilometres (42,650 feet), this stream does not have a great influence on the weather, flowing above an area of fairly consistent high pressure.

Low-level streams

Although they are weaker than higher-level streams – slow-coaches blowing at a mere 130kph (80mph) or so – these can channel air northwards and southwards over several hundred kilometres and can have a dramatic effect on the weather. One such stream, forming at 4–4.5 kilometres (approx. 13,000–15,000 feet) over Africa below the Equatorial Jet Stream, and in the same summer months, makes its presence felt in the weather of the Caribbean and southern Atlantic. Huge thunderstorms can develop around it that move outwards across the ocean. If conditions are right and the sea is a toasty 27°C (81°F) or more, hurricanes can form.

Jetting around the world

- Wasaburo Ooishi, a Japanese meteorologist, was the first to suspect the presence of jet streams back in the 1920s, when he sent weather balloons up into the atmosphere near Mount Fuji to track winds at these higher levels.
- In 1934, American pilot Wiley Post took to the sky in a specially designed pressure suit that enabled him to fly at high altitudes. Way up in the heavens, at 15 kilometres (50,000 feet), he encountered a jet stream.
- The first usage of the term 'jet stream' was in 1939, when a German meteorologist, H. Seilkopf, coined the name in a research paper.

ON YOUR TAIL

During World War II, pilots flying between the United States and the United Kingdom noticed that the flight going east was quicker than that going west, aided by tailwinds of more than 100 miles an hour. In 1952, a commercial Pan American flight from Tokyo to Honolulu, travelling at an altitude of 7.6 kilometres (25,000 feet), reduced its journey time from eighteen hours to eleven and a half hours. Since then, aircraft have exploited jet streams, not only to shorten flight times but also to reduce fuel consumption.

SECTION 4: THE AERIAL VAPOUR CYCLE

11. What Goes Around…

✍

YOU KNOW WHAT it's like… you've just spent hours blow-drying your hair to celebstandard sleekness. Then you step outside and your silky tresses morph into a candyfloss frizz. Why? It's because the air is damp and human hair stretches when it's moist. The air around us is full of varying degrees of water vapour, and it's the behaviour of this invisible vapour that is responsible for much of our weather.

Watery world

It's not for nothing that Earth is nicknamed the Blue Planet, since so much of its surface is covered with water. When it's swooshing around in the oceans, seas, lakes and rivers, water exists in the form of a liquid. But when water is warmed by the sun, it turns into water vapour (a gas that accounts for about a third of all atmospheric gases) and rises into the air. So, the world's water doesn't disappear, it is perpetually changing form and keeps circulating around the planet's weather system in a perfect example of natural recycling – and the engine that powers it all is the sun.

1. Vapour trail

More than two-thirds of the Earth's water is found in the oceans and seas, but lakes and rivers also contain large amounts. The sun acts on these wet surfaces in the same way as it does on dry land: it heats the surface, triggering a chain reaction. In the case of land, the warmed surface reflects heat, warming the air above it and causing it to rise; in the case of seas and oceans, the warmed surface waters evaporate and rise as water vapour.

Other, smaller sources of atmospheric water vapour are the moisture 'breathed out' by vegetation (in a process known as 'transpiration') and given off during the burning of fossil fuels, and in the gases that the Earth belches out during volcanic eruptions.

For obvious reasons, the Earth's wettest air is above the seas and oceans, but nowhere is the air completely dry – even above the middle of the Sahara there is some moisture in the atmosphere. It is dry simply because the hotter the heat of the sun, the faster the rate of evaporation will be.

2. Bubble, bubble…

So there's all this water vapour floating around with the other gases in the atmospheric soup that makes up what we call 'air'. What happens next? As the sun's rays heat the surface of the Earth, it in turn heats the air in contact with it. Bubbles of warmed air start to rise and as they do they also expand. It uses up energy to do this and so the bubble starts to cool down as it continues to rise. If the cooling air is full of moisture as it rises it eventually reaches a point where it is completely saturated with water vapour. The bubble rises till further cooling leads to the water vapour condensing into tiny water droplets and a cloud starts to form. What aids this process are the minute particles of dust, salt and smoke that occur naturally in the atmosphere. These provide tiny nuclei around which the droplets can condense – without this kind of miniature space debris, there would probably be many fewer clouds.

3. Transport system

It makes sense that the air above the oceans and seas would be the wettest – why, then, aren't clouds and rain confined to these places? Answer – the air itself does not stay in one spot. Moisture-laden air masses are constantly on the move, with jet streams – the fastest-moving winds of the lot – blowing clouds and weather systems about. Meteorologists have a posh word for this process: 'transportation'.

4. Waterfall

Inside the clouds the water droplets collide and join together to form larger drops. If they get large enough and heavy enough, they fall to Earth as rain. In fact, much of the rain that we get starts life as snow. It's pretty chilly up there at the top of clouds, and the tiny water droplets they contain in their upper reaches freeze into ice particles. The ice particles merge to form snow-flakes, which flutter down towards the planet's surface. If the flakes pass through warmer air on the way down, they melt – and turn into raindrops. But if it stays cold, they may remain frozen and descend as snow, sleet or hail. Rain, drizzle, snow, sleet and hail are all different forms of 'precipitation'.

5. Round and round we go

The water that has fallen from the sky eventually makes its way back to the sea. If the ground is absorbent enough, the water may soak into the ground and, when it hits impermeable rock, start to build up, spread sideways and eventually come to the surface, running downhill into streams and rivers which carry it back to the seas and oceans – for the whole process to start again. Or, if the ground is too hard or precipitation has been too heavy for the surface to absorb it, the water may run directly into the streams and rivers.

DO IT YOURSELF!

Apart from the effect it has on your hair, air humidity can also be measured by looking at trees. The leaves on oak or maple trees curl up in high humidity, giving you an early warning of rain ahead, while pine cones stay closed in high humidity and open in low humidity.

NB A temperature change of around 11°C (51.8°F) will allow air to contain approximately twice as much water vapour. So, for example, air warmed to 21°C (70°F) may be twice as saturated with water vapour as air at only 10°C (50°F), which may have seen some water vapour condense to water droplets.

Sucking it up

Around 10 per cent of the water that falls to Earth does not make its way back into the atmosphere from the sea, but from vegetation. Plants soak up the water in the ground through their roots. The moisture travels up through their stems and evaporates out into the air from pores in their leaves – creating a suction effect that draws up more water from the soil. Clever, don't you think?

It's all relative

It's behind the frizzy-hair effect – 'humidity' is the name for the amount of moisture in the air. 'Relative humidity' is the ratio (expressed as a percentage) of water vapour present in the air compared to the maximum that could be held by air at a particular temperature.

- At the top end, a relative humidity of 100 per cent is as bad as it gets – think equatorial rainforest. The air is saturated and cannot get any damper. Moisture on surfaces cannot evaporate so your skin will feel clammy and you'll have no energy.
- If relative humidity is low, the air is drier. In the inland areas of some deserts, for example, the figure can be less than 3 per cent. In such places, perspiration can evaporate more quickly so you'll feel more comfortable. Rainforest or desert – the temperature may be similar but it's the humidity that makes the difference.

WATCH YOUR LANGUAGE!

The hydrologic cycle The swanky name for the water cycle, which is made up of evaporation, condensation, transportation, transpiration and precipitation – and back to evaporation again. So now you know.

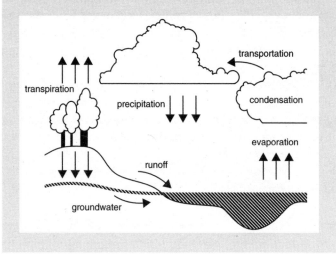

Dew see?

Another way that meteorologists measure the level of humidity in the air is by means of the dewpoint. Remember how warm air can accommodate more water vapour than cool air? The dewpoint, then, is the tipping point between saturation and condensation – the temperature at which an air mass gets too cool to hold its vapour content and this condenses into droplets.

Would you believe it?

- At any one time, the atmosphere contains enough moisture to produce about 2.5 centimetres (1 inch) of rain across the Earth's surface.
- In one year, the Earth receives a total of about 5,000 million million tonnes of rain, sleet, snow and hail.
- Most raindrops measure about 1 millimetre (0.05 inch) in diameter.
- The average raindrop contains a million times more water than the average cloud droplet.
- Race you! Raindrops can fall at up to 9 metres (291/2 feet) per second, in contrast to leisurely snowflakes, which drift down at a mere metre (3 feet) per second, at most.

12. Cloudwatch

❧

THE CLOSEST ANY of us come to the clouds is when we fly in an aeroplane. We may literally travel through them and experience the strange sensation of being totally without visibility – or we may rise into the sunlit air above, and see them spreading out below us like vast swathes of cotton wool. But clouds are more than just fascinating aerial features: they can tell us a lot about what the weather is doing.

The three Cs

The great shifting masses that we call clouds are made up of countless tiny water droplets or ice crystals, or a mixture of both, that occur when air cools to a certain point. They form according to the three Cs: condensation, convection and convergence. A moist air mass may cool when it passes over an area of cold land or sea, to create low-lying cloud or fog. Or it may cool as it rises into the atmosphere because:

- it has been warmed from below, by the warm surface of the Earth – a process known as 'convection' – and so it floats upwards, cooling as it moves further from the surface.
- it is forced upwards by such physical obstacles as mountains.
- it meets a mass of colder, denser air and rises above it.
- it is forced into a restricted area – a process known as 'convergence' – and must rise to escape.

When air reaches a certain coolness, the water vapour – which is a gas – condenses back into a liquid, and tiny water droplets form. As the droplets accumulate, a cloud forms. If the temperature is low enough, the droplets freeze into crystals of ice.

WATCH YOUR LANGUAGE!

Orographic clouds Clouds that form when air is forced to rise above mountains or hills.

Security blanket

The term 'a blanket of cloud' is not far off – clouds do protect the Earth in a similar way to a nice, cosy blanket, or perhaps a thick layer of loft insulation. They work to insulate the planet from excessive heat loss, so if the night sky is cloudy, there is less likely to be a frost. They also reflect radiation from the sun back out into space, thus regulating the amount of solar energy absorbed by the Earth's atmosphere. Clouds are also not as moisture-laden as we think. An average cloud that is 1,000 metres (3,281 feet) high can produce only around 1 millimetre (0.04 inches) of rain.

DO IT YOURSELF!

Reading the clouds

Clouds may be divided into three main categories. They've been given Latin names, which is useful because, if you know the translations, you can get a good idea of what each group looks like (see overleaf for clues)...

- *Cumulus* means 'heap' (as in 'accumulate') so these are the great piles of cloud that look like mounds of cotton wool, with a flat base.

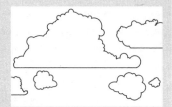

- *Cirrus* translates as 'a curl' (of hair), so these clouds look feathery or hair-like.

- *Stratus* translates as 'layer', so these are low lying clouds that may take the form of sheets of cloud without much definition – the ones we get on overcast days. Fog is a form of stratus cloud.

A further category was once used and is still found in combination with other cloud names:

- *Nimbus* means rain-bearing.

Now is when the fun begins. Simply by tacking on the appropriate prefix or suffix to the main group names, you can mix and match to describe a whole range of sub-groups of cloud. To help identify the different types, begin by looking at where they are in the sky – low down, at medium height or high up.

Lying low

Low-lying clouds begin below 2,000 metres (6,500 feet) above the ground and are usually made up of water droplets.

- *Stratus* form a grey, almost featureless blanket of cloud that is often accompanied by a light drizzle or snow.
- *Cumulus* clouds are a sign of thermals – columns of warm air rising from the ground – as you can see from the tower-like shapes at the top. They usually occur in fine weather.
- *Stratocumulus* look like separate rolls of grey or whitish cloud that form a sheet across the sky, and bring only light precipitation, if any.
- *Cumulonimbus* are the really dramatic ones – storm clouds that appear as white billows at the top, with dark, ragged, ominous-looking bases. Reaching high into the sky, they contain ice crystals as well as water droplets and are harbingers of rain, hail, snow, thunderstorms and, frequently, squally winds.

The middle way

Clouds lying at medium-height usually have a base between 2,000 and 7,000 metres (6,500–23,000 feet) above the surface and are formed of a mixture of water droplets and ice crystals.

- *Altocumulus* get their name from 'alto', which means high. They look a bit like balls of cotton wool, with dark shading and patches of sky in-between. If there is any precipitation, it rarely reaches the ground.
- *Altostratus* resemble stratus but are found higher in the sky, and create a hazy effect around the sun.
- *Nimbostratus* are the ones you'll need your umbrella for – this dark grey, ragged layer of cloud hides the sun and brings almost continuous rain, sleet or snow.

High-wire act

Starting between 5,500 and 14,000 metres (18,000–46,000 feet) above the ground, high-level clouds are usually made up totally of ice crystals.

- *Cirrus* look like wispy white threads (although they may look grey if seen against the light). They are made up of falling ice crystals, but none of these reach the ground. Dense patches of cirrus clouds may obscure the sun.
- *Cirrocumulus* look like tiny tufts or ripples high in the atmosphere, often referred to as a 'mackerel sky' because they resemble the markings on that particular fish. Precipitation from these clouds rarely reaches the ground.
- *Cirrostratus* are perhaps the most subtle of all the types of clouds – they are visible as a milky veil of cloud, so thin that they don't obscure the sun and aren't accompanied by any precipitation.

13. Not the Foggiest: Fog and Mist

∽

WHEN YOU WALK through a fog, you literally have your head in the clouds – fog is a type of stratus cloud that is so low-lying it's found at ground level. Mist is similar to fog but less dense. Both are made up of minute droplets of water, but in mist the droplets are more widely spread.

Looking ahead

On land, a layer of low-lying cloud is described as 'fog' when visibility is less than 200 metres (656 feet), while in 'dense fog' you wouldn't be able to see further than 50 metres (164 feet) ahead. Ships and aeroplanes, with more sophisticated navigation systems than the naked eye, are more tolerant of fog. Consequently, in the air and at sea, it's only called 'fog' when visibility is less than 1,000 metres (3,280 feet), and 'dense fog' when less than 200 metres (656 feet).

Fog and mist often form at night but evaporate during the day, with the warmth of the sun. On peering out of their windows at dawn, many country-dwellers will have glimpsed the ethereal sight of a layer of low-lying mist that has collected in hollows and valley bottoms – but which disperses as the sun rises and warms the ground. In winter, of course, when temperatures are lower, fog and mist may not disperse so easily.

CAUGHT SHORT

In 1815, after the Duke of Wellington, the British commander, finally crushed the French emperor Napoleon on a battlefield in Belgium, a message of victory was beamed across the English Channel, using powerful light beams in a similar way to sending Morse Code by signal lamp. Unfortunately, due to poor weather conditions, the message did not get through as intended and panic swept the nation, causing the Board of Trade to collapse. What should have read 'Wellington defeated Napoleon at Waterloo' said only 'Wellington defeated[…]'. The crucial end of the message had been obscured by fog.

A murky affair

Until the clean air acts of the 1960s, London was notorious for its fogs. It had already become noticeable back in the second half of the eighteenth century, when the smoke and noxious fumes belched out by the furnaces of the Industrial Revolution started to mix with the city's naturally damp air.

But it was with the population explosion of the nineteenth century that matters really degenerated, with chimney smoke and fog combining to create the 'smog' that would become known as the 'peasouper', a type of fog, so named because it was so impenetrable. It was smelly too, with the sulphurous odour of rotten eggs.

Peasoupers were not only murky in terms of reduced visibility; they became associated with murky acts too. In 1811, in a fog so dense that observers said you could not see more than

60 centimetres (2 feet) ahead of you, the Marr family were found murdered in the East End of London, followed, twelve days later, by the Williamson family in the same part of the city. Murder, the East End and the cover that peasoupers afforded killers became associated in public imagination.

In 1888, this culminated in a horrible fascination with the serial murders committed by the so-called Jack the Ripper. But sometimes the fog alone was the killer. In December 1952, a deadly smog descended on London. It lasted for 69 hours, leading to the deaths of more than 4,000 people.

DO IT YOURSELF!

Here's how you can tell different types of fog and mist apart...

Radiation fog occurs on still, clear nights when the ground loses the heat it has absorbed from the sun during the day, and cools. This in turn cools the air above and the water vapour it contains condenses into droplets, forming fog.

Advection fog appears when mild, damp air moves over cold ground. The lower layers quickly cool to the temperature at which condensation takes place and fog forms.

Hill fog occurs when warm, moist air is swept up onto hill- or mountaintops where the temperatures are lower. The air cools and 'hill fog' forms. In hilly or mountainous areas like Scotland, this can result in a blanket of fog that completely obscures the peak but stops abruptly halfway down the slope.

Coastal fog and sea mist occur when warm, moist air travelling across the sea cools and a resulting mist or fog forms (cold sea can have the same effect as cold land). This may then be blown shorewards by the wind.

Steam fog is brought about when rainwater falling onto warm ground evaporates and the air above gets so damp it can no longer hold any more water vapour, and the excess moisture condenses into a steam-like mist.

Freezing fog is made up of extremely cold water droplets that freeze on contact with solid surfaces such as lamp-posts or overhead cables.

14. Right as Rain:
The Mechanics of Rain

∽

THE RAINDROPS THAT keep falling on your head begin life in the world's seas and oceans, journey up into the sky in the form of water vapour, then condense back into water again and fall back to Earth. It's a long round trip and it takes a molecule of water vapour about a week to complete this cycle, during which time it may have been blown many thousands of miles. So when you next experience a downpour or a bit of drizzle, spare a thought for how far those drops have travelled to water your garden or bounce off the top of your umbrella.

'Anyone who says sunshine brings happiness has never danced in the rain.'

Anonymous

Strength in numbers

Have a look at your fingertip. Now imagine 100–200 droplets of water crowding that space – that's the density you'll find in a developing raincloud. These droplets are infinitesimal in size,

85

narrower than a human hair, and therefore aren't big or heavy enough to fall as rain. If they are to do that, what they need is strength in numbers. As these tiny droplets drift down through the cloud, they bump into each other and merge into larger, heavier droplets. With added size, weight and momentum, they fall faster and coalesce into even bigger drops until – bingo – they become a shower of rain.

Icy start

As well as water droplets, clouds may contain ice crystals. Both droplets and crystals need a nucleus to form around. Water can condense around a range of different particles including atmospheric salt and soot, but ice has more limited options: it can only form on a few types of nuclei, such as dust, with the result that there are fewer crystals than droplets in a cloud. But ice crystals make up for this by growing faster than droplets, and soon become snowflakes. If, as these snowflakes fall, they pass through warmer air, they melt and turn into rain.

Falling flat on their face

Raindrops are actually bun-shaped ovals that fall flat-side down – it's the visual trail caused by the speed of their falling that tricks our eyes into seeing them as teardrop-shaped.

World coverage

It has been calculated that if the total amount of rainfall that the whole world gets in a year were averaged out across the surface of the planet, each area would get a dose of around 800 millimetres (31.5 inches) of rain annually.

DO IT YOURSELF!

Provided it hasn't rained during the night you can have a go at predicting rainfall by taking a look at grass first thing in the morning. If the grass is dewy, it is unlikely that you'll be needing your umbrella, but if it's dry this suggests either it's going to be a bit breezy or that you'll need to throw on your rain macintosh.

Rainbows

One thing guaranteed to raise a smile when the rain starts to pour on a sunny day is the appearance of one of the weather's most beautiful visual effects – a rainbow. Some of the sun's light is refracted into its component colours in the drops of rain, producing the effect of a multi-coloured arc. The stunning phenomenon has many myths and religious stories attached to it – including the rainbow as a symbol signifying the end of the Flood in the biblical tale of Noah's Ark, and the legend of the crock of gold sitting at its end.

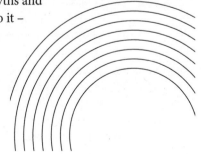

RAIN, RAIN GO AWAY

After attending church one Sunday, American author Mark Twain (1835–1910) and fellow writer William Dean Howells (1837–1920) were caught in a heavy downpour. Howells looked up at the glowering sky: 'Do you think it will stop?' he asked. 'It always has,' answered Twain.

Other types of rain

Drizzle is a kind of precipitation produced by shallow, low-lying clouds that don't have the substance or height necessary to generate a good shower. They may be less than 300 metres (1,000 feet) thick, and the misty drops they release are much smaller than full-sized raindrops.

Freezing rain occurs when snow falls through a layer of warm air and melts, then falls through a colder layer but does not freeze again because it has no nuclei to freeze onto – until it hits the ground, that is, creating a thin and treacherous layer of slippery ice that can be a real hazard to drivers.

Virga is the name for those strange, streaky curtains of rain that you sometimes see suspended in mid-air below a cloud. What's happening here is that dry air between the cloud and the ground has caused the rain to evaporate before it hits the Earth.

WATCH YOUR LANGUAGE!

Supercooled water Extremely cold water that has not frozen even though the surrounding temperature is below freezing point. Water can remain in a liquid state in temperatures as low as -40°C (-40°F).

Rainy cities

- The famous song 'April in Paris', written by the Russian-American composer Vernon Duke (1903–69), extols the charms of this romantic city in spring. Inspired by its lyrics, a friend of his visited Paris in April but had a horrible time – it poured with rain. On returning home, the friend complained to the composer. 'Why did you go to Paris in April?' Duke asked, 'The weather's always terrible then.' 'I went because of your song!' the friend cried. 'Ah,' said Duke, 'well I really meant May, but I needed two syllables.'
- London is known for its rain, but the winter of 1961 was especially bad, being the wettest on record at that time. 'The tanned appearance of many Londoners is not sunburn,' declared the *London Evening Standard*, 'it is rust.'

15. Snowed Under:
Winter Wonderland

IT CAN TRANSFORM even the ugliest urban landscape into a sparkling wonderland and bring out the child in all of us – and at the same time it can create travel chaos and make the functioning of everyday life that much more difficult. Snow is one of Nature's marvels and yet it's really no more than frozen water.

The shape of a flake

A snowflake starts life as a minute ice crystal in a cloud. It grows by attracting water vapour that freezes directly onto it, without condensing into water first. Now for the fun part: the surrounding temperature affects the way the vapour freezes onto the crystal, and hence determines the shape of the snowflake.

- Needle or plate shapes form if the surrounding air is close to freezing.
- *Dendrites* – feathery, six-branched flakes – need cooler temperatures, as low as -18°C (0°F).

- Hexagonal plates, that are less feathery than dendrites, develop if the temperature is around -28°C (-18°F).
- *Graupel* are rounded snow pellets, rather like small, light-weight hailstones, that are less flaky than other types of snow. They develop when ice crystals collide with supercooled water droplets (still liquid below freezing point because they have no nuclei to freeze onto).

Snow man

The first person to describe the structure of hexagonal snow-flakes was the Swedish archbishop Olaus Magnus, in 1555. In 1931, American farmer and meteorologist Wilson W. Bentley (1865–1931) published his book *Snow Crystals*, featuring 2,000 of the 5,000 photographs he had taken of snowflakes through a microscope.

KEEP IT PC

Snow gives everyone the chance to enjoy that traditional bit of fun: building a snowman. But, as it happens, in 2008 the record for the tallest one went to a snow*woman*. Created in Bethel, Maine, she was named after Olympia Snowe, the state's senator. Standing a proud 37.2 metres (122 feet) tall, she had eyelashes made from skis, lips from red-painted tyres, and two whole Christmas trees for arms.

All together now!

If snowflakes bump into each other as they fall, they can form monster flakes – some as big as 5 centimetres (2 inches) across. However, in the United States – where things are often done on a grand scale – even the snow may be larger than life. In January 1887, a rancher at Fort Keogh in Montana claimed to have seen a snowflake that was 38 centimetres (15 inches) wide and 20 centimetres (8 inches) thick.

White out?

Everyone knows that snow is white, right? Well, actually, they're wrong. This icy substance can have subtly different colourings. Algae – microscopic plant life – within the snow can tint it pink, blue or green, while pollen from certain conifers can even turn it yellow.

DO IT YOURSELF!

For the amateur forecaster, without the benefit of handy forecasting gizmos, your best bet for predicting snowfall is to take a close look at the skies above you. The sky tends to turn a pale grey, with lower clouds than you would expect with rain clouds. Also keep a close eye on what the birds are up to – they tend to gather close together on branches and power lines, in preparation for the snowfall ahead.

Warm enough for snow

It doesn't have to be *that* cold to snow. In summer thunderstorms, it can be snowing in the upper reaches of the clouds but mild down on the ground. And as long it falls fast enough to build up a thick enough covering, snow can settle on relatively warm surfaces on which it would otherwise melt.

Freak falls

Even in milder climes, we expect snow on mountaintops – but on the beach? Snow does not always know its place – nowhere, it seems, is safe…

- In February 1895, 51 centimetres (20 inches) of snow fell on Houston and 20 centimetres (8 inches) on New Orleans.
- In January 1977, it snowed on Miami Beach.
- In December 1997, parts of Guadalajara in Mexico received 41 centimetres (16 inches) of the white stuff.
- January 2022 saw the latest snowfall in the Sahara Desert, and in Jerusalem heavy snow shut roads, schools and businesses.

Other places, though, get what they expect – or worse…

- In February 1927, the snow on top of Mount Ibuki in Japan measured 11.82 metres (39 feet) deep – a world record.
- In 1998–99, Mount Baker in Washington State, America, achieved a world record for winter snow, with a total fall of 29 metres (95 feet) for the season.

WATCH YOUR LANGUAGE!

Sleet In Western Europe, sleet is a mixture of partially thawed snowflakes and rain, but in the United States what is known as 'sleet' is more like hail, a rarer beast than the European variety.

16. Hail and Hearty

❧

FOR THOSE OF us living in more temperate climes, the occasional light hailstorm may elicit no more than surprise. But make no mistake – those lumps of ice that fall from the heavens can be a real hazard, shattering windscreens, causing damage to property and even leading to loss of life.

How hail forms

Hailstones start innocently enough as small ice crystals in cumulonimbus storm clouds. Unimaginably vast, these clouds can have a base at 600 metres (approx. 2,000 feet) but a top stretching to a height of 12,000 metres (40,000 feet) above ground, or even 18,000 metres (60,000 feet) in tropical areas.

In the turbulent inner world of the cumulonimbus, vigorous up- and downdraughts gather the tiny falling crystals and sweep them up, down and back up again. Inside the cloud, too, are supercooled water droplets that, despite the freezing temperatures, have not turned to ice because they have had nothing to adhere to in order to freeze – yet. Now's their chance. As the droplets collide with the swirling ice crystals, they flow over them to form a new layer of ice. And they can do this again and again, creating bigger and bigger hailstones each time.

Generally, the larger the cumulonimbus, the stronger the up-draughts and hence the bigger the hail. When the hailstones get too heavy to be swept up again, they finally fall to the ground.

DO IT YOURSELF!

It is possible to observe the formation of a hailstone by counting the number of layers of ice. Each layer represents a collision with a supercooled droplet. The opacity of each layer also tells you how quickly it froze: transparent ice means it froze slowly; opaque ice means it froze quickly, trapping air to make it less see-through.

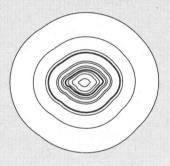

Tropical hail

In order to form, hail needs the strong updraughts and spin within a thundercloud so it is more common in parts of the world that are prone to severe storms. The largest storm clouds develop in the tropics so it is here that the biggest hailstones – often as large as golf balls – develop.

HIT THE GROUND RUNNING...

Hailstones are no slowcoaches – they can hit the ground at more than 161kph (100mph).

Hail hazard

It's no joke – hail can seriously damage your health. In some cases, it can even kill.

- In April 1986, hailstones weighing 1kg (2lb) each battered the Gopalganj district of Bangladesh, resulting in more than 90 deaths.
- In May 1996, a hailstorm in China killed around 100 people, injured 9,000 more and destroyed 35,000 homes.
- In April 1999, hailstones more than 8 centimetres (3 inches) wide clattered down on Sydney, Australia. For five hours and
- more the storm raged, damaging 40,000 vehicles and 20,000 buildings and other structures.
- In July 2010, a hailstone 20 centimetres (7.87 inches) wide fell on Vivian, South Dakota.
- In the United States, hailstorms cause more damage than any other type of weather apart from severe storms, resulting in losses to agriculture alone of about $1 billion per year.

17. The Big Freeze: Ice and Frost

༄

I T CAN LOOK as pretty as sugar frosting on candy and dazzle like a layer of crushed diamonds on rooftops – but it can also kill the plants in your garden and make it dangerous to drive: yes, it's good old Jack Frost and it's another example of what happens to moisture in the air when temperatures drop.

Going down

By day the Earth's surface, heated by the sun, acts like an electric hotplate to warm the air above it. After sundown the reverse happens as the surface radiates this solar heat back out into the atmosphere and becomes, instead, a bit like a cooling panel in a fridge. If there is no cosy blanket of cloud to insulate the Earth, the heat loss is even greater. Sometimes the temperature of the air just above the ground may even fall below 0°C (32°F) – the freezing point of water – while the temperature of the air higher up is above freezing.

When water vapour in the air settles on any ice-cold surfaces, such as the soil, or car rooftops or windscreens, it turns into ice crystals and becomes what is known as frost. But there does need to be a certain amount of moisture in the air to begin with. If the air is dry, even air temperatures below 0°C may not have enough condensing power to convert what little moisture there is into a covering of frost.

DID YOU KNOW?

Frost is white because the ice crystals of which it is made contain air.

Don't you believe it!

When meteorologists talk about 'surface temperature', don't take them at their word. Although it may sound like it, they are not talking about the temperature at ground level. Because the temperatures of air varies at different levels, forecasters take their official readings at a height of around 1.25 metres (4 feet) *above* the ground. Somewhat confusingly, they then call this 'surface temperature' and it's the figure you'll hear on the weather forecast. If on cold, calm nights the ground is colder than the air above – rather than the common daytime pattern of warmer ground and cooler air – they have a swanky name for it: 'surface temperature inversion'. So now you know.

WATCH YOUR LANGUAGE!

Frost-point The temperature at which frost forms.

What's the score?

Slight or severe, unsurprisingly, frost varies depending on how cold it gets:

- Temperatures of 0°C to -3.5°C (32°–25.7°F) bring a slight frost.
- Getting decidedly chilly: lows of -3.5°C to -6.6°C (25.7°– 20.1°F) bring a moderate frost.
- At -6.6°C to -11.5°C (20.1°–11.3°F) frost is severe.
- When it plummets to -11.5°C (11.3°F) and below, you can expect a hard frost.

Wind affects the formation of frost, too. A strong wind can slow down night-time cooling and so prevent frost developing – but if the temperature falls below freezing, the wind can make matters worse by helping the frost to penetrate.

NEVER GIVE IN

When Winston Churchill was told about an elderly man aged more than seventy-five who had propositioned a young woman in sub-zero temperatures in London's Hyde Park, he was impressed. 'Over seventy-five and below zero!' the great man exclaimed. 'Makes you proud to be an Englishman!'

Naming names

Different conditions produce different kinds of frost. If you're a gardener, it's important to know these so you can prepare yourself for what's coming. If you're a driver, it can be important too because you'll need to allow extra time to scrape the ice off your car windows before you head off in the morning – and you might have to drive extra carefully too if ice has frozen onto the roads. For farmers, frost can be a real disaster. Back in 1971, for example, it was responsible for $1.1 billion's worth of damage to agricultural crops in the U.S.

DO IT YOURSELF!

Here is how to tell different frosts apart...

Air frost This is the term that meteorologists use when the surface temperature (i.e. the air temperature at about 1.5 metres/5 feet above the ground) falls below freezing – in other words, it's extremely chilly air. Occasionally there can be an air frost when the ground temperature is above freezing, for example, in the autumn when the soil still holds some of its summer heat.

Ground frost When the temperature at 5 centimetres (2 inches) above the ground falls below freezing point, a ground frost may occur. It may not be that cold higher up, but it's what's happening low down that matters with this type of frost.

Hoar frost The most common variety, hoar frost is a layer of delicate, fluffy ice crystals that settle on grass, leaves, branches and soil or other hard surfaces such as brickwork and rooftops. It occurs on calm, clear nights when water vapour touches icy-cold surfaces and instantly freezes onto them.

Rime frost As wind blows damp, freezing air over branches, foliage and other surfaces, the supercooled droplets within the air hit these surfaces and instantly turn to ice. But because they freeze while still in motion, they can form extraordinary shapes – rather like freezing a frame in a movie sequence of someone running. Rime is rarer than the other types of frost, and the temperature needs to be extremely low for it to form.

Fern frost This is the really pretty stuff. It starts when dew forms on cold glass, then freezes. As more ice crystals form on top, a fern- or leaf-like pattern begins to emerge – hence the name.

SECTION 5: CLIMATE AND GEOGRAPHY

18. Looking at Climate

∽

YOU LIVE IN a place where the summer sun could frazzle you to a crisp and you dream of rain and drifting mountain mists. Or you live in a place where every sunny day is cause to crack open the champagne and you long for that persistent blanket of cloud to part so you can get some daylight. If you don't like the weather where you live, you need to face the truth: you are dealing with something bigger than your immediate surroundings – you are dealing with climate. And if you really can't stand it any longer, the only thing to do is emigrate.

Weather vs climate

The term 'weather' refers to variations in atmospheric conditions on a day-to-day basis – for example, whether it's cloudy, windy or rainy. 'Climate' involves a much longer view and describes the weather patterns of a particular region over a more extended period. Latitude (how far a region is from the equator), altitude (how high it is above sea level), the nature of the topography (natural features such as mountains), and distance from the sea are some of the factors in the complex mix of influences that produce different climates.

Mr Köppen's classification

In 1900, a Russian-German climatologist by the name of Wladimir Köppen tried to sort through the confusion by dividing the world's climates up into just five basic types, each with its own letter:

A. Moist tropical climates with high temperatures and high rainfall.
B. Dry climates with low rainfall and wide variations in daily temperatures – the kind of climate you would find in deserts or steppes (vast, normally flat and arid stretches of land).
C. Climates with warm, dry summers and cold, wet winters.
D. Climates typical of the interiors of large continents, with moderate rainfall and widely different seasonal temperatures.
E. Climates typical of regions of permanent ice and tundra (freezing, largely barren plains), where temperatures rise above freezing for only about four months of the year.

DO IT YOURSELF!

The Köppen system is just one of several ways of classifying climate – for example, it may also be classified into 'Tropical', 'Dry', 'Warm Temperate', 'Cold Temperate' and 'Cold'. Work out which climate you live in by checking the following definitions…

Tropical A band of territories running from the equator out to 15° north and south of it have a tropical climate. These include Central America, northeastern South America, equatorial Africa, India, Southeast Asia and northern Australia. Temperatures here are high, rainfall is heavy, and it's humid and often cloudy, with a tendency to experience hurricanes. In regions closer to the equator, the sun is nearly overhead so it's evenly hot all year. Further away, the seasons are markedly different, with wet summers and dry winters. Regions with a tropical climate are home to the world's lush rainforests or, in East Africa, to savannah (tropical grassland).

Dry The world's deserts and semi-desert regions have a dry climate. Rainfall is minimal at best and temperatures are extreme. They may be exceptionally hot, notably in the Sahara and Arabian deserts, but fall to freezing  or below at night or during the winter, as in the deserts of central Asia and western China.

Warm Temperate This type of climate brings hot summers and mild winters, with most of the rain falling either in summer – as in eastern China and the southeastern states of the United States – or winter – as in Chile, California, Western Australia and the Cape region of South Africa. There may be thunderstorms in summer but snow and frost are rare and found only in mountainous regions.

Cold Temperate Much of northwestern Europe, coastal North America and New Zealand's South Island have a maritime version of this type of climate, characterized by changeable weather, rain all year, cold winters but no great extremes of temperature. The continental version brings warm summers and cold winters and affects the interiors of large continents and landmasses, such as the plains of Russia and Ukraine and the grasslands of Canada and United States.

Cold This type of climate is found in Antarctica, Greenland and within the Arctic Circle, with temperatures near or below freezing all year and with snowfall and ice all year too. Further from the Poles, a slightly milder version of the cold climate is found, in regions such as northern Russia and Siberia and large parts of Canada. Here, summers may be hot but short, with long, bitterly cold winters.

19. Some Like it Hot: Microclimates

∽

BIG-CITY DWELLERS love the buzz of the urban jungle while country inhabitants prefer the peace and greenery of more open landscapes. But there's more than merely a lifestyle choice separating the two habitats – big cities can actually create their own mini-climates that are different from those of the surrounding countryside.

City hotspots

Climatologists have a name for this phenomenon – they call it 'microclimate', in other words, the prevailing weather conditions that apply only to a specific small area. But in the case of city microclimates, there's a more evocative name: 'urban heat islands'. Cities can literally be hotspots in a cooler rural landscape. The temperature differences can be massive: London, for example, can be approximately a staggering 10°C warmer than the surrounding countryside. And there's more: research at NASA suggests that some cities are so hot that they can even create their own weather – including violent storms with thunder and lightning.

So how does this happen? Think about it – all those concrete roads and buildings, all those brick walls, all those pavements, soak up the sun's heat during the day, then release it slowly after sundown like giant electric night storage heaters. Narrow streets and high buildings can trap heat too, giving it less space to escape; pavements cover bare ground and there is relatively little vegetation to absorb water so that when it rains, the water has nowhere to go and there is a greater risk of localized flooding. No wonder cities are warmer and more humid. In the countryside, by contrast, there are open spaces and trees, streams and rivers, which help heat to dissipate as moisture evaporates from leaves and water surfaces, and rainfall can soak into the ground.

In your own backyard

Another surprising manmade microclimate can be your own garden. In cool temperate climates, gardeners have long understood how erecting walls and fences can make their plots warmer and more sheltered, allowing them to raise those specimens that might otherwise not survive in that particular environment. Training tender fruit such as peaches or apricots against a sunny brick wall is an old trick – the brick absorbs the sun's heat and radiates it out to keep the tree warm and help the fruit to ripen. Fences can help to break the passage of prevailing winds to give additional shelter.

Hot and stuffy

Nature can create her own heat islands too, in dense forestation. The canopy of a thick forest forms an almost unbroken surface that soaks up the sun's heat during the day, in a similar way to the surface of the ground. This warm canopy conserves the heat but shades the forest below so that the lower you go, the cooler it is: the difference may be as much as 5°C (41°F) during summer. In addition the trees block the free flow of air through the forest. So within the forest it may not only be cooler than the surrounding countryside, but more humid too.

20. The Earth's Spin: How it Causes Seasons

∽

WE MAY NOT like it but we are in fact spinning in space on an outsized lump of rock, a.k.a. Planet Earth. Not only that, but the lump isn't even upright – it's whirling crookedly, at an angle. It's this wonky tilt that is the cause of our seasons and the weather conditions that come with them, and the length of our days and nights…

Get the picture

Imagine a rod running through the centre of the Earth, between the North and South Poles. This is the planet's axis and it is tilted at 23.5°. The planet spins around this axis while at the same time orbiting the sun – a journey of 365 days, or a year. The Earth's tilt means that different parts of its surface are angled towards the sun at different stages in its orbit, while other parts

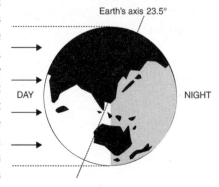

Earth's axis 23.5°

DAY

NIGHT

are angled away. The parts angled towards the sun get more hours of sunlight while the parts angled away get correspondingly less. The result? Varying lengths of day and night and changing seasons.

How the seasons happen

When the Northern Hemisphere is tilted towards the sun, the solar rays hit it more directly, giving more intense heat – and creating summertime. At the same time, the Southern Hemisphere is angled away from the sun, so the sunlight has further to travel and it is wintertime there. As the Earth travels on in its orbit, the South starts to point towards the sun and the North away from it, reversing the seasons.

DO IT YOURSELF!

Although global warming is effecting their movements, the flight path of geese is still a fairly good indicator of change in the seasons. Generally you will notice geese flying together in a 'V' shape, heading south when colder weather is on its way in winter, and will return when more temperate seasons come round.

Another good indicator of seasonal change that nature gives us is when the temperature of the soil is once again warm enough for plants to grow (at around 16°C or 60°F).

How night and day happens

As the Earth orbits the sun, it spins around on its axis at the same time. Different sections of its surface revolve to face the sun in turn – and it is this that creates night and day. But the planet's tilted axis makes it a bit more complicated than that. Although a region may have turned to face the sun so that it's technically daytime, it will receive different amounts of sunlight depending on whether it is angled towards the sun or away from it in that particular period – and hence the varying length of days and nights in different parts of the world in different seasons.

The most extreme examples of this phenomenon are the Poles. If you were standing on, say, the North Pole in the polar summer when the Arctic is tilted towards the sun, you would enjoy twenty-four-hour daylight – for six months of the year. Conversely, in the Antarctic winter when the South Pole is angled away from the sun, there would be no daylight at all – just six months of darkness. Then there is the equator, which is the pivotal point of the planet's axis – rather like the centre point of a seesaw. Here, the tilt is balanced out, giving roughly equal lengths of sunlight in summer and winter, and days and nights of about twelve hours each.

LOUSY SUMMER

The coldest winter I ever spent was a summer in San Francisco.

Attributed to Mark Twain

Look on the bright side

Sunshine is associated with happiness and after the grey skies of winter nothing cheers like a bright spring day – and there's a scientific basis for this. It begins in the hypothalamus, a control centre in the brain that is responsible for regulating such important bodily functions as mood, energy, sleep and sex drive. When sunlight passes through the retinas of our eyes, it stimulates the hypothalamus to release serotonin – a feel-good chemical – and slow down the production of melatonin, the hormone that makes us sleepy. In winter when sunlight levels are lower, we may experience SAD (seasonal affective disorder) and suffer from sleep problems, fatigue, depression, anxiety, irritability, overeating and loss of libido – and all because of a lack of adequate sunlight.

Hot enough for you?

As well as lack of sunlight, the human body – and brain – struggles when the weather gets too hot and humid. In heatwaves, people:

- are more irritable and lethargic, find it harder to concentrate and have difficulty sleeping.
- turn to crime – excessively hot, sticky weather leads to irrational behaviour; in New York, for example, the regular summer crime waves are thought to be linked to the high temperatures.
- are more likely to die, especially if they are elderly – when the thermometer hits 38°C (100.4°F) or more, mortality rates increase by about 10 per cent.

Something in the air...

Persistent, noisy, seasonal winds such as the mistral of southern France and the föhn of the Alps are another weather feature that can drive us humans mad, making us feel more tired, anxious and depressed – and, according to studies, more likely to have traffic accidents, commit crime or attempt suicide. Such winds can lead to temperature increases of as much as 15°C (59°F) in just a couple of hours. The air in them is also positively charged which is not conducive to an upbeat mood. We respond much better to negative ions – atoms or molecules with a negative electrical charge – which explains why we usually feel better after a heavy downpour when the air is full of negative ions.

21. The Lie of the Land: Climate and Mountains

∽

IF THE FACE of the Earth were flat, the weather would be much the same across large stretches of land. But with hills and mountains making the planet's surface decidedly bumpy, the flow of weather-bearing winds is interrupted, resulting in strongly contrasting conditions on either side of the opposing slopes.

In the rain shadow

Mountains get in the way of weather. When moisture-laden air hits a mountainside, it has nowhere to go but up. As it rises, it cools and the water vapour within it condenses, forming clouds that release much of their watery load in the form of rain or – if it's cold enough – snow. The result? Wet conditions with plenty of rain on one side of the range, dry weather with little rain on the other. These dry regions on the far side are said to be in the mountains' 'rain shadow'.

Take the Himalayas, for example. These giants are made up of several parallel ranges, including the Greater Himalayas that soar to the giddy heights of 6,000 metres and more, and contain the world's highest peak, Mount Everest, at 8,849 metres. As the moist summer winds blow up from the south, the Himalayas literally stop them in their tracks. Blocked by these massive barriers of rock, the winds are forced upwards, cooling as they go and forming clouds that eventually burst in the summer monsoon rains. On the plateau north of the Himalayas, however, it's a different story – here the landscape is arid and desert-like.

The way the wind blows

The planet's prevailing wind patterns cause more rain to fall on the western sides of mountains in the middle latitudes, and on the eastern sides in the tropics. This is what happens with the Rocky Mountains in the United States. The prevailing winds, blowing in from the west, give Washington state, on the western side of the Rockies, plenty of rain – while the states of the northern plains, to the east of the mountains, largely have to go without.

DO IT YOURSELF!

Mountains can create their own clouds too, when air rises up and over the peak and down the other side. If the air cools enough on its upward journey, it may form a particular type of cloud that can take different shapes. Here's how you can tell what kind of clouds have formed:

- *A lenticular cloud* looks a bit like a flying saucer and hovers some distance from the peak.

- *A cap cloud* engulfs the top of the mountain.

- *A banner cloud* streams from the peak like a flag –
  banner clouds often form at the top of the Matterhorn
  and the Rock of Gibraltar.

These clouds do not move because they are not blown by
the wind, but form with the upward and downward passage
of air.

The white stuff

Because temperature drops with altitude, there can be snow on
mountains near the equator while it's baking hot down at ground
level. Climb 300 metres up a mountainside and it's estimated that
you'd experience the same cooling effect as you would if travelling 350 miles towards the poles at sea level.

WATCH YOUR LANGUAGE!

Orographic Used to describe any weather feature created
by mountains or hills, such as clouds, rain or wind. Go on,
blind friends and family with science as you speak impres-
sively of 'orographic effects'.

Wind tunnel

As well as forcing winds to blow up and over them, mountain
chains can also act as funnels, forcing winds through the valleys

between them – which makes the winds stronger. One well-known example is the mistral of southern France, which roars along as it is squeezed through the Rhône Valley between the Alps and the Massif Central.

22. Current Account: Ocean Currents

WARM OCEAN CURRENTS are part of Nature's central heating system – like the water-filled radiators in your home, they heat the air around them and can change regional climates, making them warmer than they should be for their latitude. Cooler currents, in contrast, work like an air conditioning system to bring down temperatures in places that would otherwise be much hotter.

Moderating influence

Nearly three-quarters of the Earth's surface is covered by water so it's hardly surprising that the sea has such a big influence on climate and weather. Just like the water in your central heating radiators, the sea warms up and cools down in a slow, measured way. In autumn, it gradually radiates the heat it has stored from the summer so that areas closest to it, along coastlines, are milder than those further inland. In summer, the process works the other way. While the sea slowly warms up after winter, it produces cooling breezes that give coastal regions lower temperatures than those inland.

Going with the flow

Propelled by the prevailing winds, the biggest ocean currents flow in clockwise loops in the Northern Hemisphere and anti-clockwise in the Southern Hemisphere – and, as they swirl about, they transfer heat from the tropics northwards to the polar regions. For example, warm water flowing towards the North Pole gives northwestern Europe and western Canada milder winters than they have a right to, given their distance from the equator. Conversely, cooled water flowing back from the polar regions produces cooler-than-expected summers in the coastal regions of Spain, Morocco, Namibia, California, Ecuador and Chile.

Keep moving!

The effect that ocean currents can have on climate is perhaps most powerfully demonstrated by the Gulf Stream, one of the world's strongest currents. It all works rather like the way in which air masses move, with warm air rising and cold air sinking and winds rushing from areas of high pressure to areas of low pressure, to even out the differences. Here's what happens…

- Way up in the North Atlantic, winds blowing from the Arctic cool the surface water, causing it to become colder, denser and more salty.
- This cold, dense, salty water sinks to the bottom of the ocean and begins to flow towards the equator, where it will gradually warm up.
- Meanwhile, surface water is warmed in the Gulf of Mexico. As the cold water heads south, this warm water moves northwards into the North Atlantic to replace it – and the whole circular movement of currents is repeated.

- The northward current of warm water is known as the Gulf Stream – and it is responsible for northwestern Europe having an average annual temperature that is 9°C (48.2°F) higher than it should be for its latitude.

Hot and cold

Two ocean currents give marked differences in climate – and sea temperature – to the coastal regions of South Africa:

- Flowing south from the tropics, the benevolent Agulhas current warms the country's eastern shores and makes the ocean invitingly mild for swimmers.

- Flowing north from the Southern Ocean around Antarctica, the freezing Benguela current cools South Africa's western shores and makes for a surprisingly icy dip for unsuspecting tourists in the Atlantic waters off Cape Town.

WATCH YOUR LANGUAGE!

Continentality The big differences in temperature variation in inland areas of continents, as compared with coastal regions. Because of their proximity to the sea and its moderating influence, coasts experience less variation in temperature.

23. The Hottest and Coldest Places on Earth

∾

SOME PLACES ON Earth are so hot – or so cold – that it's impossible to imagine how any living creature could survive there. When it comes to extreme temperatures, the following places are record-breakers:

What a scorcher!

… and the contenders for Hottest Place on Earth are:

- The *Gandom Beriyan* – meaning 'Toasted Wheat' – region of the Lut Desert in Iran. In 2021, land surface temperatures of 81°C (177°F) were measured here and in the Sonoran Desert of North America – that's hot enough to fry an egg. In fact, the region gets so hot that it defeats the survival skills of even the most primitive life forms – not even bacteria have been found here.
- *El Azizia* in Libya – on 13 September 1922, it achieved a high air temperature of around 57.8°C (136°F).
- *Death Valley* in California – this rivals El Azizia with occasional air temperature highs of around 56.6°C (134°F).
- *Dallol* in Ethiopia – it may not be a gold-medal winner, but Dallol remains a reliable contender for Hottest Place on Earth with average annual temperatures of 34.6°C (94°F).

Brrr...

When it comes to the prize for Coldest Place on Earth, there is really only one winner:

- *Vostok* in Antarctica – on 21 July 1983, the air temperature plummeted to -89°C (-128°F). It hardly bears thinking about. But it isn't always that bad. Sometimes Vostok gets positively balmy, with temperatures of -25°C to -19°C (-13°F to -2.2°F) – it's enough to make you want to put on your shorts.

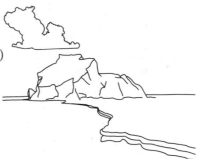

NB The South Pole comes in at a close second in winter, with average lows of around -60°C (-76°F).

Hottest and coldest

Some places want to win all the prizes...

- *Verkoyansk* in Siberia. This contender for both Hottest and Coldest Place on Earth boasts winter temperatures of as low as -70°C (-94°F) while summer highs may soar to 38°C (100°F).

NB Deserts experience extremes of day and night-time temperatures, with the burning desert sun in the daytime easily heating sand or rocks to 60°–70°C (140°–158°F). But at night, all that heat dissipates into the atmosphere and temperatures plunge to below freezing.

Dry as a bone

If you can't stand the constant damp and drizzle where you live and want to move to a nice, dry climate, here are a couple of options:

- Parts of the Atacama Desert in Chile, which can get as little as 1 millimetre (0.04 inches) of rain per year. At that rate, it would take a century to fill a coffee cup.
- Antarctica – there is, of course, a lot of water in Antarctica but it's all locked up in the ice sheet. With an average thickness of 2,160 metres (7,087 feet) the ice sheet contains around three-quarters of the world's fresh water. If it were to melt, average sea levels would rise by about 58 metres (190 feet). But when it comes to rain, Antarctica receives very little – less than 5 centimetres (2 inches) of precipation per year, which is about the same as the Sahara and makes Antarctica technically a desert. In the interior, the region known as the Dry Valleys has not received rain, it is estimated, in two million years.

Wet, wet, wet

If you don't want to drown, you should avoid:

- Lloro in Colombia. This receives an average of 1,330 centimetres (524 inches) of rain each year – or, to put it another way, that's about 13 metres (43 feet).

SECTION 6:
EXTREME WEATHER

24. Weathering the Storm

∽

L OVE 'EM OR hate 'em, no one would deny that thunderstorms are among the weather's greatest showstoppers. Most people have experienced such events, except perhaps for the inhabitants of areas where it hardly ever rains, such as the Sahara and the polar regions. Here, as Eliza Doolittle might have said, thunderstorms hardly happen.

Opposites attract

Thunderstorms start with the formation of cumulonimbus clouds – thunderclouds generated by the rapid rise of masses of warm, moist air. As these violent updraughts whoosh to the top of the cloud in the freezing upper reaches of the atmosphere around 5 kilometres (3 miles) above the ground, they are matched by violent downdraughts of icy air heading towards the bottom.

Cumulonimbus clouds are highly electrically charged. How this happens isn't fully understood but it may be caused by the collisions between minute ice crystals inside the cloud.

The turbulence of the air masses leads to huge differences in the electrical distribution within a cumulonimbus, with a positive charge tending to collect at the top, and a negative charge at the bottom – the ground is also positively charged. Because opposites attract, when these differences build up sufficiently, an electrical discharge or current shoots between the negatively and positively charged areas. We most commonly see this as cloud to ground lightning, but the discharge may also occur within a cloud or between clouds.

WHAT A SHOCK!

In June 1752, the American scientist Benjamin Franklin (1706–90) performed a legendary test to prove his theory that lightning was a form of electricity. He knew that metal attracts electricity so he chose a key as his conductor. Franklin attached the key to one end of a kite string and tied the other end of the string to an insulating silk ribbon that he could wind around his hand. He then flew the kite in a storm and, when a spark flashed on the key, he knew his theory was correct. This was a highly dangerous and potentially fatal thing to do, and the only witness was his son.

Joining the dots

Think of aerial photographs of traffic at night – they look like lines of light tracking the path each vehicle has taken. Or cartoon films in which the individual frames flick by so fast that the human eye sees them as continuous movement. Lightning's a bit like that too.

What we perceive as fork lightning is the path of the electrical sparks caused by electrons as they move from one place to another. But the electrons are travelling so fast that we see these sparks as lines rather than individual flashes of light. In the case of fork lightning, what we actually see is not the faint 'leader' stroke coming down from the cloud, but the much brighter return stroke that shoots back up from the ground.

The big bang

Lightning's electrical discharge generates huge amounts of energy and heats the surrounding air to unimaginably high temperatures, causing it to expand violently. A second or two later, the air cools and contracts again. It's this atmospheric boom-and-bust that creates the massive shock wave of sound that we call 'thunder'.

LOUSY WEATHER

John Piper (1903–92) was an English painter who specialized in dark, moody, stormy landscapes. When Britain's King George VI (1895–1952) was shown some of his work, he commiserated with the artist: 'Pity you had such bloody awful weather.'

Don't you believe it…

… lightning *can* strike twice! And if you need proof, take a look at these facts:

- In a single storm, New York's Empire State Building – one of the tallest buildings in the city – attracted 15 lightning strikes in about as many minutes.

- Roy Cleveland Sullivan, a park ranger in Shenandoah National Park, Virginia, USA, was struck by lightning seven times between 1942 and 1977. The strikes cost him the nail on his big toe, knocked him unconscious, singed his eyebrows and burnt his hair (twice) among other injuries.

But don't worry – experts have calculated that your chances of being hit by lightning are about one in three million.

Enough to make your hair stand on end

- At any one time, there are about 1,800 thunderstorms occurring somewhere in the world, producing 100 bolts of lightning per second.
- A bolt of lightning can travel a distance of 96 kilometres (60 miles), racing along at 322,000kph (200,000mph).

- In a clear night sky, you can see lightning up to 160 kilometres (100 miles) away, but you're unlikely to hear thunder more than 30 kilometres (19 miles) away.
- The average lightning strike is 3.2–14.5 kilometres (2–10 miles) long and carries 100 million volts at a current of 10,000 amps. As a comparison, 240 volts at a current of 0.1 amps is enough to kill an adult.
- In less than a second, lightning hits the ground with 300,000 volts of electricity and heats the surrounding air to 30,000°C (54,000°F) – that's five times hotter than the temperature on the surface of the sun.

TROPICAL STORMS

In the tropics, thunderclouds can tower up to 18,000 metres (60,000 feet) above the surface of the Earth. All that turbulent activity up in the sky means that tropical storms are frequent and often violent. The Democratic Republic of Congo has been estimated to have the greatest number of thunderstorms in the world, with more lightning strikes than any other country.

What doesn't kill us...

… makes us stronger, or so the German philosopher Friedrich Nietzsche (1844–1900) would have it. Not necessarily true. Lightning kills only about 10 per cent of its victims, usually through cardiac arrest, but if it doesn't do this, it can give you some extremely nasty experiences. For example, it can:

- knock you off your feet.
- make you jump (by causing your muscles to contract).

- make your hair stand on end.
- give you amnesia.
- cause burns.
- make you (temporarily) deaf or give you ringing in your ears (tinnitus).

It has even been claimed to be able to strip your clothes off.

DO IT YOURSELF!

Because light travels faster than sound, you'll see lightning before you hear the accompanying thunder. Light whizzes along at 300,000 kilometres (approx. 186,000 miles) per second, while sound dawdles along at a paltry 330 metres (1,080 feet) per second. This means that you'll see the flash of lightning before you hear the rumble of the accompanying thunder – and you can use this snippet of information to work out how far you are from a storm…

1. Count the number of seconds between the lightning flash and the thunder.
2. Divide by five and you'll get the distance in miles. So, for example, a time difference of ten seconds means that the storm is two miles away.

If you prefer your measurements in metric, a time lapse of 3 seconds represents a distance of 1 kilometre – or you can count 300 metres for every second.

DANGEROUS PURSUITS

If you want to minimize your chances of being hit by lightning, give up golf and angling. These two hobbies are said to be two of the more risky outdoor pursuits during storms.

Keeping safe in a storm

Lightning usually strikes at the beginning or end of a storm and this can be up to 16 kilometres (10 miles) from its source, in places where the sky is still sunny and clear. It may also be hidden by clouds. Here's how to keep safe in a storm:

Do...

- Put down any implement, especially anything made of metal, such as a golf club, fishing rod, etc., and move away from it.
- Count the seconds between a lightning flash and a thunder roll. If this is less than thirty seconds, the lightning is dangerously close, so seek shelter at once.
- Choose a suitable shelter. Though no shelter is totally safe, the typical house offers reasonable protection, as does an enclosed metal vehicle such as a car or bus.
- Close the windows.
- Stay out of open, exposed places, such as fields.
- If you are caught outside in a storm, go into the crouch position, put your feet together and squat down, tucking your head into your knees.
- Spread out if you are in a group outdoors, leaving at least 6 metres (20 feet) between you.

- Stay away from prominent objects outdoors, such as tall, isolated trees, poles or towers. If you are near trees, smaller ones are safer.
- Get out of the water – the swimming pool, the sea, the boating lake. If you are a keen sailor, ensure that any boat you step into is adequately earthed.
- Get medical attention as quickly as possible for anyone who has been struck by lightning. Don't delay.

Don't…
- Use your phone, watch television or use your computer in a storm. Lightning seeks the easiest route to earth – and wires, cables and water pipes provide ideal conduits for it, increasing the risk of lightning striking your home.
- Lean against car doors or touch the steering wheel, ignition or gear stick as these can act as lightning conductors.
- Rely on rubber to protect you – the soles on your shoes or the tyres on your car won't completely insulate you from lightning shocks.
- Go outside to watch the storm.

WATCH YOUR LANGUAGE!

Ball lightning This appears as a slow, moving ball of light which explodes or burns out. And no one knows what causes it… cue spine-chilling music.

Blue Jet A phenomenon observed by satellite above large thunderclouds – a rapid, blue, conical-shaped flash of light above the centre of a storm.

Fork lightning An electrical discharge between the clouds and the ground. Its jagged shape is due to the way the electricity is finding the easiest path towards the ground. It is one of the two most common forms of lightning, along with sheet lightning.

Red Sprite A burst of red light above thunderclouds, similar to blue jet.

St Elmo's fire A rare phenomenon consisting of a continuous glowing discharge from pointed objects such as masts, spires and trees.

Sheet lightning An electrical discharge within a cloud or between two clouds, it looks like a blinding flash of light illuminating the sky.

25. The Eye of the Hurricane

❧

'IN HERTFORD, HEREFORD and Hampshire, hurricanes hardly happen,' sang Eliza Doolittle in the musical *My Fair Lady*. Lucky old Hertford, Hereford and Hampshire. Hurricanes can cause an enormous amount of damage, including loss of life.

How do hurricanes happen?

Hurricanes are vast, swirling tropical storms rotating around areas of extremely low pressure. From edge to edge, a whole hurricane storm system is alarmingly big – up to 16 kilometres (10 miles) high, 800 kilometres (500 miles) wide and travelling along like a spinning top at up to 64kph (40mph). So far, no one knows exactly what turns a tropical storm into a hurricane, but the following factors contribute:

- Warm, moist air – hurricanes start over the sea, and only over water with a surface temperature of more than 27°C (80.6°F). They form within the latitudes of 5 degrees and 30 degrees, where the surface waters are more likely to achieve such temperatures.
- Sufficient spin from the Earth's rotation – this contributes to the rotation of the hurricane.

WATCH YOUR LANGUAGE!

Hurricane A tropical storm over the Atlantic in which the average wind speed exceeds 119kph (74mph). Given the right atmospheric conditions, a hurricane can develop in as little as twelve hours, and can be hard to predict.

Tropical cyclone Same as a hurricane, but occurring over the Indian Ocean and affecting regions such as the Bay of Bengal and Bangladesh, as well as East Africa, Indonesia and the northern coasts of Australia. (Yes, you remembered right – *cyclone* is also a general term for a low-pressure centre, without the nasty wind.)

Typhoon This also the same as a hurricane, but occurring in the western Pacific and affecting the Phillippines, China, Vietnam and Japan.

Because of the way the planet spins, hurricanes swirl anti-clockwise in the Northern Hemisphere and clockwise in the Southern.

This is how the experts believe hurricanes happen:

1. The warm sea surface heats the air above, causing a current of hot, damp air to rise rapidly and leaving a centre of low pressure behind at the surface.
2. This low-pressure centre attracts winds that spiral inwards, in an attempt to fill the low-pressure void.
3. Warmed by the sea, these incoming winds rise and create another area of low pressure, which in turn draws in more fast-moving air. And so a self-sustaining energy loop forms.

4. Higher up in the atmosphere, the hot, moist air cools and condenses to form massive cumulus and cumulonimbus clouds.
5. At the same time, the Earth's spin causes the current of rising air to twist until it forms a whirling cylinder of cloud around a core of cloud-free, relatively still air – the 'eye of the storm'.

Wind and wave

The most common phenomenon that people associate with hurricanes is an extremely strong wind. Hurricane winds are powerful, it is true, and have been known to reach speeds of up to 320kph (200mph). But there are other potentially devastating features of hurricanes, such as waves up to 15 metres (50 feet) high and ocean swells, or storm surges, when the level of the ocean rises in response to the storm. Both can cause widespread flooding. And then, of course, there is the downpour: each day, a hurricane picks up around 2 billion tons of moisture and later releases it in the form of rain.

A journey of 1,000 miles...

When a hurricane heads inland, it can cause horrific destruction. But as the storm travels over land or moves into an area of colder water it will no longer be getting the heat and humidity that feeds it. The friction of moving over a land surface also alters the airflow, and as a result, the system collapses – the eye of the storm fills with cloud and the hurricane dies. However, don't be complacent: if heat and humidity remain high enough, a hurricane system may have a lifespan of two to three weeks. The average hurricane covers a distance of 5,000 kilometres (3,000 miles), at a daily rate of 500–650 kilometres (300–400 miles).

Keep your hat on

The time to be on the alert for hurricanes depends whether you live in the Northern or Southern Hemisphere.

- In the Northern Hemisphere – in the Atlantic and in the eastern and western Pacific – the time to hold onto your hat is between July and October.
- In the Southern Hemisphere – in the Indian Ocean and off Australia – the danger period is between November and March.

Can you see it coming?

Well, no. Other than awareness of the general global pattern of hurricane formation, meteorologists currently have no way of predicting particular hurricanes before they arise. Once such a storm has flared up, however, it is possible to track its path, using such aids as satellites, radar and reinforced aircraft with measuring equipment on board. This at least allows forecasters to give the public early warning of what's on the way.

To describe the severity of a hurricane, many meteorologists use the Saffir-Simpson Hurricane Scale (or SSHS for short, as it is a bit of a mouthful). This was developed in the United States in 1969 by Herbert Saffir, a structural engineer, and Bob Simpson, director of the National Hurricane Center. The scale measures hurricanes in terms of their wind speed and power to devastate, ranging from Category 1 (gusts of 119–153kph/74–95mph, causing damage to trees and mobile homes) to Category 5 (winds in excess of 251kph/156mph, causing severe damage even to the most substantial buildings and requiring wide-scale evacuation of the population).

However, the SSHS is used only to measure hurricanes in the Atlantic and the eastern North Pacific. In other parts of the world, different scales are employed.

As easy as A, B, C

Because there can be as many as 100 hurricanes causing havoc in different parts of the world every year, meteorologists needed a way to distinguish between them, and the naming system (or systems) was born. In the Caribbean in the early days, hurricanes were named after the saint's day on which they occurred. So the storm that hit Puerto Rico on 13 September 1876 – Saint Philip's Day – was called 'Hurricane San Felipe'. During World War II, meteorologists with the US air force and navy gave hurricanes the same names as their wives and other female loved ones. After this, it began to get more complicated. Today, hurricanes are named according to the letters of the alphabet and according to which storm zone or 'basin' they belong to, as well as alternating between male and female names.

THE GREAT STORM

The BBC's weatherman Michael Fish will go down in history as the forecaster who got it horribly wrong. On 15 October 1987, a mystery female caller phoned the Corporation with some worrying news. Later, in his evening television forecast, Fish referred to the call. 'Earlier today,' he said, 'a woman rang the BBC and said she had heard that there was a hurricane on the way. Well, if you are watching, don't worry,' he soothed, 'there isn't.'

A few hours later while the nation was in bed, the south-east of England was hit by the worst storm since 1703. Winds reaching 185kph (115mph) ripped up 15 million trees and brought some crashing down on cars and houses, tore roofs off buildings, brought down power lines, reduced the famous Shanklin Pier on the Isle of Wight to a pile of drift-wood, blew a Channel ferry aground at Folkestone in Kent and led to the deaths of eighteen people.

26. Tornado Twisters

They may not be as fast as a speeding bullet but they do outstrip hurricanes in velocity: with potential top speeds of around 480kph (300mph), tornadoes are the fastest winds at the Earth's surface.

Super storms

Snaking across the landscape, a tornado looks like an ominous biblical pillar of smoke. In fact, tornadoes are violently rotating columns of air that form below cumulonimbus storm clouds and reach all the way to the ground. Dust, sucked up into the swirling air, makes them clearly visible. Tornadoes do not occur in isolation but need a parent thunderstorm of a particularly severe type – known as a supercell – in order to develop.

Supercells form when cold air masses from the poles meet very moist, warm air masses from the tropics, resulting in unstable atmospheric conditions. If there is sufficient turbulence, the wind increases, air is funnelled into a rotating

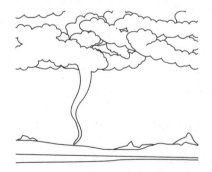

column and a tornado begins to form. Tornadoes may also occasionally form within hurricanes, for example, in the Gulf of Mexico region and in the southeast United States.

Let's do the twist

Simply observing the size of the funnel does not provide enough information to assess a tornado's strength, and in 1971 a new measure was devised to rate tornadoes according to wind speed and the amount of damage they could do to manmade structures. Now widely used by meteorologists, the Fujita-Pearson Scale was developed by Japanese researcher Ted Fujita, of the University of Chicago, and Allen Pearson, head of the National Severe Storms Forecast Center.

The Scale has five categories, from F0 to F5. The weakest tornadoes – which, fortunately, are the most common – belong to the first F0 group. A mere 5–15 metres (16–50 feet) wide and with wind speeds of between 64 and 116kph (40 and 72mph), these so-called 'Gale Tornadoes' may cause light damage to chimneys and branches, and may blow down shallow-rooted trees. At the other end of the Scale, F5 or 'Incredible Tornadoes' may be 1.6–4.8 kilometres (1–3 miles) in width, with wind speeds of 420–512kph (261–318 mph). They can destroy houses, throw vehicles considerable distances and even damage steelreinforced concrete buildings.

Fact or fiction?

In Frank L. Baum's famous story *The Wizard of Oz*, which begins in Kansas, a tornado starts the storyline, picking up Dorothy, her dog Toto and her entire house, which it then dumps on the Wicked Witch of the East, crushing her to death. The tornado's ability to cause such mayhem is not far from the truth of reallife tornadoes. They have been known to throw trains off their tracks, transport pieces of paper more than 160 kilometres (100 miles) and take people on magic-carpet rides while still ensconced in the bathtubs or on the mattresses where they were when the tornado struck.

Tornado Alley

Many parts of the world – even unlikely places not normally associated with violent weather, such as the Netherlands and the United Kingdom – experience tornadoes. But some areas are more at risk. As well as certain atmospheric conditions, the geographical make-up of a particular region can make it more prone to the kind of storms that produce tornadoes. One such infamous region is the central zone of the United States nick-named Tornado Alley, which includes the states of Texas, Oklahoma and Kansas. Here, on 3 May 1999, a record tornado struck northern Oklahoma – it was 1.6 kilometres (1 mile) wide with a wind speed of more than 480kph (300mph). But Tornado Alley isn't the only part of the world to suffer – way over on the other side of the planet, the Ganges Valley in northeastern India and Bangladesh also experiences deadly tornadoes that are responsible for numerous fatalities.

Take shelter!

If you are unlucky enough to find yourself in a tornado, be sensible and follow these safety tips:

- Take refuge in a solidly constructed building, such as one made of brick, not wood.
- Head to the cellar or basement, if there is one. If not, seek out a small, interior room without windows on the lowest floor, such as an internal bathroom.
- Tuck yourself under a table and cover yourself with a blanket or cushions to protect yourself from flying debris.
- Stay out of mobile homes, vehicles or large, open rooms.

Water feature

When a tornado forms over large rivers, lakes or oceans, sucking up vapour into a spinning column, it is known as a water spout or sea spout. Typical spouts span no more than 75 metres (246 feet) and involve gale- or hurricane-force winds of up to 240kph (150mph). But they are short-lived phenomena, lasting a mere twenty to thirty minutes.

27. High and Dry: Droughts

❧

A BARE LANDSCAPE, cracked earth, the bleached carcasses of dead cattle and desperate, hungry people – these are television images of drought that are beamed into our homes and are what many people think of when they hear the word. In fact, droughts can be much less dramatic than this and can strike almost anywhere on the planet – even in countries that we think of as damp and rainy.

It's all relative

The definition of a drought depends on the climate of the country. Even the United Kingdom – a country not noted for its dry, hot weather – has its droughts. In the frequently wet UK, fifteen consecutive days with less than 0.25 millimetres (0.009 inches) of rain on any one of those days was once categorized as a drought. The definition was later abandoned as too rigid. And it could not, in any case, be applied globally – that kind of rainfall might almost amount to a deluge in the Sahara, for example. Technically, deserts like the Sahara do not suffer from drought because lack of rain is a permanent condition and therefore normal for that region.

Droughts, then, are periods that deviate from the norm for a particular region or country: there may be a longer than usual dry stretch, or rain may be noticeably lighter than normal.

As well as looking at water shortages, another way of measuring droughts is by observing the effect on crops – rather like the scales that measure hurricanes and tornadoes according to the damage they cause.

Something in the air

The amount of water vapour in the atmosphere determines how much precipitation (rain, sleet, snow and hail) there will be, and therefore whether or not there will be a drought. But winds play their part too. Winds blow air masses around so that a mass of cool, moist, oceanic air over a stretch of the Earth's surface may be replaced with a mass of warm, dry, continental air.

The temperature of the ocean can also have a dramatic effect on global weather patterns and, hence, droughts. For example, unusually warm surface waters in the Indian Ocean are thought to have contributed to droughts in the Sahel, along the southern border of the Sahara. Similarly, El Niño – a strange, recurring warming of the surface waters of the tropical eastern Pacific – can turn weather patterns topsy-turvy thousands of kilometres away, pumping heat into the air above and shifting air masses above the ocean so that wet regions become drought-prone and dry regions become wet. Conversely, El Niño's opposite number, La Niña, makes the ocean water cooler. Both El Niño and La Niña usually take one to three years to run their course.

No water, no shade

There are other, lesser factors that can contribute to drought. In places that depend on melt-water (water produced by the melting of snow or ice), a lack of snow in the winter means that there will be less water around in the spring and summer. A drought can even perpetuate itself through the conditions it creates: as vegetation dies off in the dry heat, the land surface is exposed; with less vegetation to provide shade, the bare ground absorbs and radiates more solar heat, thus helping to maintain the clear, rain- and snowcloud-free skies above.

Dry as dust

For city dwellers experiencing bans on the use of hoses for watering gardens and washing cars, a drought may be no more than an inconvenience. For farmers out in the country – especially small-scale peasant farmers – it can be a lot more serious, as they watch their crops wither and die through lack of water. But farmers have been blamed, too, for making the situation worse through poor agricultural practices that leave the topsoil bare and exposed. When drought strikes and the sun beats down relentlessly, the soil bakes hard and the upper layers dry to powder, becoming highly susceptible to erosion by wind.

This is what the farmers of the Dust Bowl discovered in the 1930s in the United States (see page 150). Agricultural conservation practices, such as planting shelter belts of trees to minimise soil erosion, can help to lessen the effects of a drought. In Niger, for example, belts of neem trees helped to reduce the impact of drought in the Sahel region of western Africa in the 1970s.

DO IT YOURSELF!

Whilst there's not much you can do to predict or indeed prevent a drought single-handedly, you can certainly help to protect your garden from the effects of drought by 'mulching'. Mulching means spreading lawn cuttings/leaves/straw/shredded newspaper etc. around your plants. This conserves soil moisture and helps control weeds, which steal much of the water needed by your plants.

The effects of droughts

Water shortages, dried-up rivers, crop failure and soil erosion… As if these weren't enough, there is an even more serious consequence of drought – wildfires. Tinder-dry vegetation, scorching temperatures and lack of rain provide the perfect conditions for such blazes that ravage natural environments and destroy lives and property. In some drought-prone countries, such as Australia, wildfires are a regular seasonal occurrence…

- In 2002 and 2003, Australia's summer brought the severest drought then on record, and sparked numerous wildfires.
- In February 2009, temperatures soaring to 40°C (104°F), roaring winds and a decade-long drought formed a deadly alliance that sparked bush fires that engulfed parts of Victoria, New South Wales and South Australia. Fatalities numbered around 180, and more than 7,000 people were made homeless.
- Summer of 2019–20 saw Australia's costliest natural disaster as bushfires killed 34 people, destroyed some 3,000 homes and devastated 24 million hectares (60 million acres).

YOU SCRATCH MY BACK...

In some parts of the world that experience regular bush fires, plant life has formed a kind of symbiotic relationship with fire and actually depends on it for its survival. A good example is the Cape Floral Kingdom region of South Africa, close to Cape Town. Here, the summer months are dry and blazes are common. Many of the seeds of fynbos plants, such as proteas, germinate only after a bush fire. In the meantime, they play a waiting game, protected within woody cones for at least one year. Fire does two things for the seeds: the flames burn off competing vegetation that would otherwise deprive the seedlings of light and the ash enriches the soil. Researchers have also discovered that chemicals in the smoke stimulate germination.

28. Sandstorms

TOGETHER, DROUGHT AND wind conspire to cause dust storms and sandstorms; that's obvious. But airborne dust and sand particles can have some unexpected – and even beneficial – effects.

Deserts and plains

We associate sandstorms with deserts – those parts of the world that are permanently drought-ridden. These include the Sahara, the Gobi Desert of Mongolia, the Taklamakan Desert of north-western China and the Thar Desert of India. Then there are other regions, such as the Great Plains of North America, that experience occasional periods of drought and become more susceptible to dust- and sandstorms at those times.

The high winds needed to shift quantities of sand and dust form when there are strong updraughts, caused by the rapid rise of hot air. Cooler air sweeps in underneath the rising hot air, and the resulting wind picks up the loose particles of sand and dust on the arid surface and carries them aloft, sometimes transporting them thousands of kilometres. The fronts of these storms may look like solid walls of sand or dust, rolling across the landscape to heights of up to 1,524 metres (5,000 feet).

SHIFTING SANDS

Enormous quantities of dust and sand can be moved from one place to another during a storm; roads, railway lines and houses can be obscured or buried; agricultural land can be stripped of its topsoil; and if the wind is strong enough, whole sand dunes can be reshaped or even blown away and the sand deposited elsewhere, to form a new dune.

THE DUST BOWL

In America during the 1930s – the Depression years – additional hardship was visited on large numbers of the American population when drought affected some 20 million hectares (50 million acres) of the Great Plains region. The dry-as-dust soil was picked up by the wind and dust storms became a regular occurrence. In an area that became known as the 'Dust Bowl', dust buried houses, farms and even whole communities and caused widespread suffering and even death. Farmers left the region to seek a living elsewhere.

Tom Joad, the central character in the classic *The Grapes of Wrath* by John Steinbeck (1902–68), epitomizes their struggle, when he and his family are forced to leave their Oklahoma farm in search of a better life in California. Bad agricultural practices by European settlers were blamed for the erosion and undoubtedly contributed. But studies have shown that this was not a one-off disaster; similar 'dust bowls' had occurred in the past, long before the arrival of the settlers in the region.

Motes of dust

There is always dust in the air. Larger particles of sand or soil that have become airborne will, eventually, be deposited back onto the Earth's surface. But tinier particles stay up there for much longer. The deserts around the Arabian Sea supply most of the Northern Hemisphere's summer dust clouds. Indian monsoon winds transport this dust towards Asia and North Africa, while dust clouds blowing westwards are swelled by Sahara sand. In the Southern Hemisphere, most of the duststorms originate in Australia's Outback.

Crossing the ocean

Gigantic clouds of dust and sand from Sahara storms travel an extraordinarily long distance and have some unexpected benefits:

• Saharan dust clouds regularly travel right across the Atlantic ocean, from the desert all the way to the rainforests of South America – a journey of around 8,000 kilometres (5,000 miles). The dust acts as an atmospheric sunscreen, bouncing heat back into space and cooling the Earth's surface. NASA research has shown that dust clouds reduce the temperature below them by 1°C (33.8°F), in a similar way to a raincloud.

- The mineral content of the dust helps to fertilize the ocean, encouraging the growth of plankton and thus supporting the marine food chain. When it reaches the South American rainforests, the dust also deposits valuable minerals there.
- The dust clouds may even help prevent hurricanes. The hurricanes that pound Florida form off the west coast of Africa. Scientists have found that if a budding West African hurricane hits dusty air from the Sahara, it reduces the risk of it developing further.

WHAT THE DEVIL?

Dust devils are in the same category as dust storms, but are smaller, shorter-lived and much less devastating. They form in clear weather, when the sun's rays hit the ground with enough heat and intensity to lead to the formation of thermals – bubbles of warm, rising air – that carry particles of dry, dusty soil up with them.

29. Open the Floodgates

❧

FLOODS HAPPEN BECAUSE of exceptionally heavy rain or storms, it's true, but that's only part of the story. The amount of rain that falls within a certain period of time is significant in causing this weather phenomenon too – and human actions can make the situation worse.

Nowhere to go

Think about it – almost everywhere in the world has rain and some places more than others, so why aren't we all drowning in endless floods? The simple answer? The water travels or changes form: some of it evaporates, some of it soaks into the ground and some runs off into streams and rivers that eventually carry it to the sea.

But it's all a matter of quantity and speed. If surface water builds up to such an extent that these natural 'overflow' mechanisms can't cope, flooding may occur. Recent weather, soil type and terrain also play a part. If the ground has already reached saturation point due to previous heavy rainfall, it may not be able to soak up more moisture, while dense, clay soils are more impenetrable to water than light, sandy soils. Mountains can add to the risk of flooding, too, by channelling rain into dangerous torrents, and rivers may become so swollen with extra water that they cannot contain it and burst their banks.

The human element

Concrete pavements and asphalt roads are good for us humans, keeping our feet free of mud and giving our cars smooth surfaces to drive over. But these impermeable materials also make it harder for water to soak into the soil, and can thus contribute to floods. Storm drains, too, if poorly maintained or – worse still – blocked, leave rainwater with nowhere to go but rising upwards.

Water wall

Flash floods are sudden floods that can occur without warning and are potentially lethal, catching people off guard and leading to fatalities. The source of such a flood may be a thunderstorm a distance away that, in a short space of time, unleashes large quantities of water that are beyond the ground's ability to soak up. In less than a minute, the water level may rise by up to a metre (3 feet) and eyewitnesses have described seeing a rushing 'wall of water'.

Rain, rain, go away

Some parts of the world are more flood-prone than others. Cherrapunji, at the foot of the Himalayas in northeast India, is one such area – and one of the wettest places on the planet. Because of the weather patterns and the mountainous terrain here, monsoon rains and tropical cyclones can produce downpours of up to 200mm (8 inches) per hour. Rains of such magnitude bring life to the earth, but they can also produce flash floods and mudslides that take life, including that of humans.

THE WETTEST PLACE ON EARTH?

- Cherrapunji receives more than 10,000 millimetres (394 inches) of rain per year. But 1861 was especially wet, with an annual total of 26,000 millimetres (1,024 inches) of rain.
- Mawsynram, in the Khasi Hills of northeast India, goes one better than its near neighbour Cherrapunji with an average total annual rainfall of 11,971 millimetres (471 inches), most of which falls in six months.
- The island of Réunion in the Indian Ocean enjoys the distinction of having perhaps the heaviest downpours ever. Over a period of twenty-four hours in March 1954, 1,850 millimetres (73 inches) of rain fell there, followed closely on 7 and 8 January 1966, with 1,825 millimetres (72 inches).
- Guadeloupe, in the Caribbean, was drenched with 38 millimetres (1½ inches) of rain in a single minute on 26 November 1970.
- Mount Wai'ale'ale in Hawaii sees almost continual rain, with only approximately two weeks' worth of dry days per year.

30. Weird Weather: Raining Cats and Dogs

∽

SWARMS OF LOCUSTS, FROGS, and gnats, downpours of hail and fire and the waters of the Nile turning to blood – the plagues visited by God on the Egyptians for refusing to free the Israelites are as nothing compared with the real thing – weird rain.

Heaven-sent

Frogs are just some of the strange objects that are reported to have fallen from the skies in the world's most peculiar showers...

- In 1841, life near Lebanon, Tennessee, in the United States, got positively grisly when a local farm was showered with blood, fat and muscle tissue.
- In February 1861, shortly after an earthquake, fish fell on Singapore.
- In 1873, a storm unleashed an aerial bombardment of thousands of frogs on the unsuspecting citizens of Kansas City, Missouri.

- In 1890 a downpour of birds' blood hit Calabria in southern Italy.
- In 1948, Bournemouth on the south coast of England had a fishy experience – a shower of herring.
- Between 1982 and 1986, Colorado had several downpours of kernels of corn.
- In June 2005, it was frogs once more. This time the small amphibians fell in their thousands on the inhabitants of Odzaci, Serbia. Some frogs survived to hop off.

Given the power of tornadoes to transport the most unlikely items over long distances, it is not inconceivable that frogs, fish, etc., were sucked up into tornadoes and sea spouts and then jettisoned later – far from their original location – when the wind died down. As for showers of blood, an explanation is harder to find. One suggestion is that violent winds tore apart flocks of birds and it was their remains that fell from the sky.

SECTION 7:
CLIMATE CHANGE

31. Global Warning

SCIENTISTS HAVE, FOR some years, been warning us that the Earth is heating up and that climate is changing. Today nearly all scientists point the finger at human activities as the major cause. In order to reduce the damage caused by climate change, many countries have agreed to a 'net zero' approach – cutting out emissions of heat-trapping or 'greenhouse' gases such as carbon dioxide. In 2015, the Intergovernmental Panel for Climate Change set up by WMO and the UN also aimed to prevent the Earth's temperature rising more than 1.5°C (34.7°F). By 2023, many scientists feared that this target would not be met. So, what does climate change mean for our weather?

Face the facts

Scientists don't mince their words. Our planet, they say, is going through a massive shift in climate not seen in the past 10,000 years. What's the evidence? Here are just a few of the figures:

- Since the pre-industrial period of the 1850s, global temperature has risen 1.2°C (34.16°F).
- Scientists estimate that by the year 2100 global temperatures will rise by 4°C (39.2°F).
- Scientists estimate that global temperatures will have risen by 1.5–6°C (34.7–42.8°F) by the year 2080.
- Over the last 100 years, global sea levels rose by about 20 centimetres (8 inches). The biggest rise happened since the year 2000.
- The warmest 8 years on record have all been since 2015: 2016, 2019 and 2020 were the hottest.
- Europe saw its hottest summer on record in 2023, with an average temperature of 16.77°C (62.19°F) during June, July and August, 0.66°C (33.19°F) above average.
- There has been a rise in precipitation; in parts of the Northern Hemisphere the increase has been 0.5–1 per cent every decade, with heavy downpours increasing by 2–4 per cent.
- Since the late 1960s, snow cover in the Northern Hemisphere has decreased by about 10 per cent.
- The amount of carbon dioxide (CO_2) – a key greenhouse gas – in the atmosphere has increased by about 50 per cent since the Industrial Revolution, and it's on the up. The oceans have absorbed so much CO_2 that their acidity has increased by 30 per cent.

MELTDOWN

Depending on the level of global warming, sea levels could rise up to a metre (3 feet) by the end of the twenty-first century.

Nature or nurture?

It is natural for the world's climate to go through cycles and undergo major fluctuations. At present we are in an interglacial period – a slightly warmer global average stretch that has continued for about 10,000 years since the end of the last ice age that spanned 100,000 years. Continental drift, the Earth's tilt and orbit and the weather pattern known as El Niño (see page 145) all contribute to the cooling or the heating of the planet's atmosphere.

Recent climate change has fluctuated too. Temperature increases in the first half of the twentieth century were followed by a drop in temperature mid-century that led scientists to worry that the Earth was cooling. It seems, though, that this cooling-off was a blip in the larger trend of global warming, which picked up again in earnest in the 1970s. Whatever the natural causes, the overwhelming evidence is that human industry and other activities are affecting the climate. We need to change our behaviour now, before it is too late.

Not seeing the wood for the trees

Planet Earth's climate could be likened to a beautifully functioning clock, with all the parts working together to ensure the smooth running of the whole. Remove or alter one of those parts, and you change the whole. Take forests, for example – they have a major impact on atmospheric conditions:

The good...

- Vegetation plays a key role in the hydrological cycle (the natural system which endlessly recycles water, changing it from liquid to vapour and back again). While humans 'perspire', trees and other forms of vegetation 'transpire' – they give off moisture from their leaves. The water evaporated in this way accounts for around 10 per cent of the water transferred to the atmosphere.

- In the process known as photosynthesis, trees (and all other plants) soak up carbon dioxide from the air and so act as filters to remove this harmful greenhouse gas from the atmosphere. At the same time, they produce oxygen.
- Trees shade the Earth's surface, letting the soil retain moisture. Also, about three-quarters of land-based species live in forests.

... the bad...

But – in the drive for more agricultural land – vast swathes of the rainforests that once covered our Green Planet have been cut down. For example:

- Rainforests once covered 14 per cent of land, but now cover only 6 per cent.
- Between 1960 and 2005, Asia lost nearly 40 per cent of its rainforests. But reforesting programmes restored 1 million hectares of forest between 2000 and 2005.
- Only around 10 per cent of West Africa's rainforests still stand.
- Burning forests is the second biggest source of greenhouse gases after burning fossil fuels.

... *and the ugly*

Clearing tropical rainforests has a number of adverse effects:

- It destroys wildlife habitat and threatens the survival of species (habitat loss is one of the biggest dangers for wildlife).
- It turns once-lush areas into arid lands, with the risk of soil erosion (because there is less vegetation to shade the ground and keep it moist, and fewer roots to hold the soil in place).
- It affects global weather patterns and rainfall, often in places a long distance apart, because the world's weather is an integrated system. For example, studies have shown that deforestation of the Amazon Basin can alter precipitation levels in Texas, while destroying the rainforests of Southeast Asia could change rainfall patterns far to the west in the Balkans.

ACID TEST

Burning fossil fuels, in vehicles and coal-fired power stations, releases sulphur dioxide and nitrous oxides into the air. Carried on the wind and mixing with the water vapour in the atmosphere, these can then produce weak sulphuric and nitric acid that falls in the form of 'acid' rain (there can be acid snow or fog too) – sometimes considerable distances from where the pollutants were first generated.

The effects are most harmful in regions where the soil or water is not naturally alkaline enough to neutralize the acid. Among the repercussions of acid rain are polluted rivers and lakes with dead fish, and forests denuded of foliage. Fortunately, many governments are now working to reduce the amount of pollutants released into the air, and some species – such as the vulnerable brown trout – are making a comeback.

Look at it this way

Climate change is a hugely complex subject that cannot be viewed in black-and-white terms – and some of its effects might surprise people:

Protective gases

The 'greenhouse' image depicts Planet Earth surrounded by a bubble of gases that trap heat from the sun, rather like the glass of a greenhouse. Without this shield of gases, a large amount of heat would escape to outer space and our world would become inhospitable to human life, with average global temperatures 30°C (86°F) lower than at present and well below freezing. However, an excessive build-up of these gases – particularly carbon dioxide from industrial processes and the burning of fossil fuels – has been blamed for adding to the 'greenhouse effect', in which Earth's atmosphere becomes too hot.

Wet and dry

Higher temperatures mean more evaporation. More water evaporating from oceans and other large bodies of water means more moisture in the air – and thus a greater potential for rain.

So, global warming may actually make certain parts of the world wetter. Conversely, if more moisture evaporates from the soil in drier areas of the world, they will become even drier. The overall result? More floods and more droughts.

Don't be too sure...

For its latitude, northwest Europe has a milder climate than it has any right to – this is because the Gulf Stream, carrying warming tropical waters up from the Gulf of Mexico, modifies the region's climate (see pages 120–1).

But over on the other side of the planet, global warming might change this. Higher temperatures in northeast Canada, Greenland and the Arctic could lead to higher rainfall (higher temperatures = more evaporation = more rain) and more ice melting. This might cool the Gulf Stream, which would drop temperatures in the UK by 3.4°C (38.12°F), reducing rainfall and damaging agriculture.

Longer seasons

The warmer temperatures and extra carbon dioxide associated with global warming could stimulate plant growth, and are increasing the length of the growing season in some countries. If other weather patterns – such as rainfall – don't change for the worse, it's possible that some parts of our planet could become more productive agriculturally.

Killer and lifesaver

Ozone – a reactive form of oxygen – is a paradoxical gas. At ground level, emissions from car engines interact with sunlight to produce ozone, which at this level is a health hazard. Up in the stratosphere, however, ozone forms a shield against the ultraviolet light from the sun, and protects us from skin cancer and other illnesses.

Chlorofluorocarbons (CFCs) – chemicals once used in such products as refrigerators, cleaning solvents, spray cans and air conditioners, destroyed stratospheric ozone, creating a hole in the layer. They were phased out under the Montreal Protocol (1987). However, it can take between 50 to 100 years for the ozone layer to recover completely, so it still needs time to be fully restored.

In the shade

While greenhouse gases warm us up, aerosols cool us down. We're not talking here about the stuff that comes in spray cans, but about minute airborne particles such as soot, dust, smoke, grit and other nasties emitted by vehicles, factories and homes. These aerosols reflect back some of the sun's rays and so – surprisingly – help to keep the atmosphere cooler. Volcanic eruptions and large fires have the same effect – when Mount Pinatubo in the Philippines erupted in 1991, it sent up a cloud of volcanic dust that blocked the sun and cooled the Earth.

However, aerosol pollutants cause respiratory and other illnesses. Also, they can impact on rains. For example, the disastrous floods in Pakistan in 2022 might have been worsened by aerosols, so many countries are reducing them. Ironically, this could speed up global warming.

MOO!

Atmospheric levels of methane, another of the key green-house gases, have also increased dramatically, doubling over the last 200 years. According to NASA scientists, around 60 per cent of methane emissions are now due to human interventions: oil and gas production; burning these fossil fuels; organic decomposition in places such as rice paddies – and by domesticated or cattle cows belching and passing wind during their digestive process. (Tropical termites are gasbags, too.)

Atmospheric methane breaks down within ten years, so the United Nations points out that if we reduce its emissions by 45 per cent, then global warming would be reduced by an important 0.3°C (32.54°F).

32. Harnessing the Elements: Alternative Power

∽

CONCERNS OVER CLIMATE change have accelerated the drive to find alternative sources of power that do not affect the Earth's atmosphere or damage its ecosystems. But 'green' power is nothing new. We human beings – inventive creatures that we are – have been harnessing the elements for our own ends for centuries.

A fair wind

More than 5,000 years ago, an observant boatsman somewhere must have realized that by catching the wind he could give his vessel an extra push, and so the sailing ship was born – and, with it, a new form of power that would serve mankind well for many centuries.

The earliest-known large sailing ships, dating from around 3,000 BC, ploughed their way around the Persian Gulf. In the fifth century AD, the Vikings took the technology to new levels and crossed the Atlantic Ocean in their sailing ships. Later, in 1492, Christopher Columbus made it to the New World with his fleet of three sailing ships – including the new-style, speedy 'Niña', based on the design of the Arab 'dhow'. But with the rise of the steamship in the nineteenth century, the great age of sail was over.

Spin the sails, turn the wheel

Wind provided another early source of green power. The new, high-tech invention of the ancient world – the windmill – automated the job of grinding grain and pumping water, saving people the bother of having to do it by hand. Although there are claims that it was invented in China more than 2,000 years ago, the first-recorded Chinese windmill dates from 1219, while the earliest-known design comes from Persia somewhere between 500 and 900 AD.

The potential of water did not escape people's notice either. The waterwheel, consisting of paddles mounted around a wheel that is turned by the force of moving water, first made its appearance thousands of years ago. Like the windmill, the water-powered wheel automated such tasks as grinding grain and pumping water for drinking and irrigation. Later, it was pressed into service to drive sawmills and other mechanized processes. In 1769, the invention of the water frame by Richard Arkwright (1732–92) – powered by a waterwheel – revolutionized the cotton industry, turning it from a cottage industry dependent on the spinning wheel to mechanized, factory-based production.

Green and clean

It is ironic that with all our technological advances we are – for the sake of our future and of our planet – turning once more to sources of power that have always existed. The forces of Nature are awesome, as electric storms and hurricanes prove, but how do we harness this free and relatively clean energy, and can it work anywhere in the world? The different forms of green energy have their advantages and disadvantages. Here are some of the options:

Wind power

An inexhaustible source of power, wind can be used to turn large turbines that generate electricity. A group of turbines together earn the name 'wind farm'.

- The electricity generated by wind farms is generally no more expensive than conventionally produced electricity.
- Off-shore farms, where the wind is stronger and more consistent, cost more to build than on-shore ones – but in a country like the UK, with large stretches of windy coastline, off-shore wind farms could easily meet the nation's needs, with power left to spare.
- Wind farms are of concern to wildlife conservationists because the rotating blades of the turbines kill bats as well as thousands of migrating birds each year. But in 2009, a Texas wind farm came up with a solution – NASA-style radar that can detect approaching flocks and shut down the turbines until the birds have passed. However, conservationists still maintain that wind farms should be located away from major migration routes.
- Other researchers point out that more birds are killed by hitting windows or buildings in general than by wind farms.
- In early 2023, wind farms in the UK generated more electricity than gas for the first time, producing a third of the nation's energy. This was an increase from just 4 per cent in 2012.

Solar power

This is the big one, at least in terms of the source of power itself – the sun. Solar power works by capturing some of the sun's energy in panels made up of photovoltaic, or photoelectric, cells. These then use the energy to generate electricity or to heat water.

- The current technology is not advanced enough for solar power to be used on a large scale, and photovoltaic cells were originally very expensive. However, their price approximately halved in recent years, so this energy source became more viable for general use. Importantly, it does not need a large-scale national grid infrastructure.
- The sector grew by 26 per cent a year from 2016 to 2022, when 15 per cent of Australia's energy was supplied by solar power.
- For obvious reasons, solar power is best suited to countries with a hot climate and many hours of strong sunshine. It is a less effective choice for regions with cool, changeable, cloudy weather.
- To increase the amount of power the panels can generate, large mirrors can be arranged to reflect more sunlight onto the cells.

Tidal power

The raw force of the sea is another inexhaustible powerhouse of renewable energy, and may become an important way to reduce carbon-based energy. One method of generating electricity from the sea shores up the water in dams or manmade lagoons, to exploit the difference in height between high and low tide. The other, more practical option makes use of the power of tidal currents:

- Tidal power is a more reliable source of energy than solar or wind because the tides are predictable and not subject to changing atmospheric conditions.
- Because it depends on the state of the tides, tidal power is an on-off source of energy, able to provide electricity for only six to twelve hours a day. If these hours do not coincide with peak demand, the electricity generated must be stored and a backup source of energy must be available for when tidal power is low.
- Tidal power stations are inexpensive to run but expensive to build, so it can take a long time to recoup the initial costs. However, prices are falling and it is possible that by 2030 tidal power will be cheaper to produce than nuclear energy.

Hydroelectric power

Humans have long exploited the force of falling water to turn waterwheels and millers' grinding stones, but hydroelectric power has gone one step further. By damming up rivers, engineers allow a powerful head of water to build up behind the dam walls; when released, the water rushes out with such force that it turns the turbines that generate electricity.

- Hydroelectric power is not subject to the ups and downs of weather and tides – it can be made available on demand, simply by releasing the pent-up water.
- Hydroelectric plants cost a lot to build, but they are cheap and easy to run. So, in the long term, they may prove to be cost-productive.
- To achieve a sufficient head of water, whole river systems may need to be dammed and this can interfere with the ecology of the region. Massive flooding on one side of the dam can mean loss of habitat for wildlife and loss of homes

for people, as well as submersion of historical sites. Meanwhile, lack of water downstream can be equally disastrous for plants, animals and people.

- Dams in hot countries can be environmental polluters. Here, in the hotter months, the water levels fall and lush vegetation rapidly sprouts from the newly exposed, damp soil. When levels rise once more, the plants are submerged and rot – releasing methane gas into the atmosphere.

Geothermal power

The ground on which we stand is, in itself, a reliable source of what's known as geothermal power. At its core, the temperature of the Earth is around 5,500°C (9,932°F) – that's as hot as the face of the sun – and the top 3 metres (10 feet) of the planet's surface remain at a consistent 10–16°C (50–61°F). Some countries have spotted the potential of this free and eco-friendly energy source, and use geothermal power to heat water, to regulate temperature in homes and workplaces in winter and summer, and to generate electricity.

- Unlike wind and solar power, geothermal energy is not dependent on changes in atmospheric or seasonal conditions – it is available all year round.
- The United States Environmental Protection Agency says that geothermal heating and cooling systems are possibly the most energy efficient and environmentally safest systems available.
- But as a renewable natural resource it is not evenly distributed across the globe. The greatest resources are found in regions where there is volcanic activity and that have natural hot springs, such as Iceland and Japan. Icelanders have been reaping the benefits of clean and inexpensive geothermal power for more than 70 years.

Biomass energy

Although we need to stop burning fossil fuels, we can burn biomass. This is organic matter, often the by-product of industry, agriculture or domestic life. Dried out, these materials can be converted to electricity, providing energy and reducing waste at the same time.

SECTION 8: THE WEATHER IN CONTEXT

33. The Bigger Picture

WHEN YOU'RE WONDERING if the rain will hold off for your barbecue, or whether your flowers will benefit from a downpour before they wilt away, or when you're pondering whether you're wearing enough layers to keep off the cold – have you ever stopped to consider the bigger picture? That weather and climate just may have a greater influence on our lives than we think, and that sometimes they have even turned the tide of human history?

I'll huff and I'll puff...

... and I'll blow your house down. The Three Little Pigs knew they had to build shelters to guard themselves from the Big Bad Wolf. In the same way, we humans have come up with ingenious methods of shelter to protect us from the elements. Be it mud hut, hide yurt, skin tepee, snow igloo, log cabin, frescoed Italian palazzo, turreted French château, Scottish island castle or modernist house, the fundamental purpose of all human buildings is the same: to keep their inhabitants comfortable and safe from the vagaries of the weather. If it weren't for wind, rain, hot sun and the freezing cold – along with a strong exhibitionist streak – would we have the Palace of Versailles?

Wrap up well

With little body hair to keep us warm, we 'naked apes' – as the naturalist Desmond Morris dubbed us – need extra help to cope with the great outdoors. The answer? Clothes! What began as skins and furs have, over the centuries, become 'fashion' and we have also come up with some pretty inventive items to beat the weather:

Macintosh A waterproof raincoat (also spelt mackintosh), named after Scottish chemist Charles Macintosh (1766–1843), who invented a rubberized fabric. The first went on sale in 1824 – no more getting soaked when caught in the rain!

Umbrella That reliable, portable canopy, which helps keep off the rain or the sun (when it is sometimes called a parasol). Derived from the Latin word *umbra*, meaning 'shade', it is of great antiquity, used in the ancient Middle East, in Egypt, ancient Greece, India and the Far East. Traditional Japanese paper parasols trace their history back to the silk ones of China. Shipwrecked on his desert island, Daniel Defoe's fictional hero Robinson Crusoe fashions his own sun umbrella from skins so he can 'walk out in the hottest of weather'.

Wellington boots Waterproof footwear based on eighteenth- and nineteenth-century Hessian boots (from Hesse in Germany), worn by the military and aristocracy and popularized by the

Duke of Wellington, after whom they are named. They are also known as gumboots.

The first Americans

During the last Ice Age, which ended about 10,000 years ago, much of the water that covered the Earth's surface was frozen in the form of glaciers, so that sea levels were approximately 100 metres (328 feet) lower than they are today. Land that would otherwise be covered by water was exposed and 'land bridges' joined different parts of the world; for example, Great Britain and France were linked, as were Australia, Tasmania and New Guinea. Before the ice retreated and water flooded in to fill what is now the Bering Straits, people crossed the land bridge that joined Asia to Alaska and became – it is believed – the first human inhabitants of North America.

The Deluge

The Sumerians had Utnapishtim and his wife, the ancient Greeks had Deucalion and his wife Pyrrha, the ancient Hebrews had Noah and his wife Naamah – all survivors of a great flood which drowned the known world and in which the rest of humanity perished. From the Middle East to the Amerindian cultures of Central and South America, people around the globe tell their own story of a cataclysmic deluge.

So widespread are these tales that it is tantalizing to speculate whether they are ancient race memories of an actual event, passed down through the generations, that morphed into myth along the way. In the case of the Mediterranean and Middle Eastern versions, this may well be true. Although the world had been warming up and glaciers had been melting after the last ice age, 6200 BC or thereabouts saw a temporary return to freezing conditions. In this mini-ice age, sea levels dropped again, exposing once-covered land. The level of the Black Sea fell by about 100 metres (328 feet) and people settled around it. Then, in about 5600 BC the global warm-up resumed and sea levels rose once more. The Mediterranean spilled beyond its shores and poured into the Black Sea, and it may be this that inspired the famous Old Testament story.

Another suggestion is that the biblical flood recalls a memory of a cloudburst in the mountains of Armenia in about 3200 BC which, it is said, caused the Tigris and Euphrates rivers to flood, covering 103,600 square kilometres (40,000 square miles) of ancient Sumerian villages and land with a 2.4-metre (8-feet) layer of mud and rubble. In the *Genesis* account, it was the flood to end all floods:

'And the rain was upon the earth forty days and forty nights… And the waters prevailed upon the earth an hundred and fifty days'.

Wind power

Think about it…

- Without the wind to fill his sails, would Columbus have reached America, or Cortés Mexico or Pizarro Peru? And if the conquistadors had not made it to the New World in their sailing ships, spurred by the smell of gold, would the native peoples have been subjugated and most of South America now be speaking Spanish?

- Without the wind to fill their sails, would Portuguese explorers have rounded the Cape of Good Hope in their search for a trade route to the lucrative spice reserves of the East? Without the wind to fill their sails, would the Dutch and English have followed, and begun the colonization of southern Africa?

An ill wind

It's an ill wind that blows nobody any good.

Traditional saying

In 1588, Philip II of Spain sent his Armada – a fleet of armed vessels that represented the might of Spain – to attack England, then ruled by Elizabeth I. Having reached the English Channel, the Armada engaged in running battles with the English fleet, which finally dispersed it with a fireship attack. It was then that the Spanish had their first skirmish with a far more fearsome enemy: the weather. Strong winds blew them northwards into the North Sea and, with their original plan of attack scuppered, they decided to sail home. But worse was to come. The latter part of the sixteenth century was characterized by exceptionally strong North Atlantic storms and, as the Armada rounded Scotland and sailed into the rough seas between it and Ireland, fierce gales relentlessly pounded the Spanish ships: only about half of the Armada survived. When Philip heard what had happened, he is said to have declared: 'I sent the Armada against men, not God's winds and waves'.

Cold snap Beginning approximately in the thirteenth century and ending in the mid-nineteenth, the Little Ice Age ushered in a period of severe winters – and had some far-reaching effects on human behaviour, including…

Witch hunts Witches were thought capable of controlling the weather so, as weather worsened, the number of witch trials rose. In his research of European prosecutions, historian Wolfgang Behringer has found a link between particularly icy weather and a rise in witch trials in the years 1560–74, 1583–89, 1623–30 and 1678–98.

On thin ice? Rivers and other waterways that do not freeze today used to turn into highways of ice. People were able to skate on them and even held 'frost fairs' on them. The first such fair on London's River Thames was in 1607.

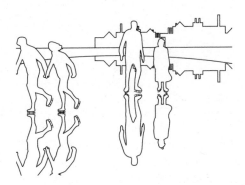

Ice bridge In 1658 the strait known as the Great Belt between the Danish islands of Zealand and Fyn froze over. This was good news for the Swedish army because they were then able to march across the ice to invade Copenhagen.

Walking on water In the winter of 1780, the water of New York harbour froze, enabling the inhabitants of the city to walk from Manhattan to Staten Island.

Drought and ice

Weather has even played a role in fomenting revolution. When anticyclonic conditions hit France in the spring and early summer of 1788, a severe drought ensued, resulting in a greatly reduced grain harvest. This pushed up the price of bread, the staple food of the workers, and made a bad situation worse: the cost of this staple had long been a focus of popular discontent and there had already been numerous bread riots.

The harsh winter that followed brought more suffering: fuel prices increased and flour mills could not operate because the water that turned their wheels was frozen. In July 1789 the cost of bread rocketed and anger over starvation, taxation and royal control erupted into mass violence. On 14 July, a mob stormed the Bastille, a fortress used as a state prison in Paris. They beheaded the Marquis de Launay, the prison commander and some of his officers, and paraded their heads around the streets. With what became known as Bastille Day, the French Revolution had begun.

34. Weather Myths and Legends

✌

IMAGINE HOW IT would be to have no understanding of the science behind the weather, no accurate method of forecasting, to be totally at the mercy of the elements for your survival. Wouldn't you then see weather phenomena as chaotic, unpredictable and destructive – and might you not then turn the elements into gods, to be revered and appeased?

Mother Earth and Father Sky

The ancient Greeks pictured the primal forces of weather and the powers that shaped our planets as a mighty family of passionate and violent gods.

The first of all the gods was Gaia, who conceived and bore a child by herself. He became Uranus, Father Sky, who was both her son and husband. Earth and Sky together brought forth numerous children including the Titans, Cyclopes and the Hecatoncheires.

Altogether the children of Earth and Sky were an unruly brood that wanted nothing more than to roam the world spreading tumult and destruction. So Uranus, their father, imprisoned them deep in the belly of their mother, Earth... But in time, the Titans escaped and an all-out war between the Olympian thunder god Zeus and his allies (including the Cyclopes and Hecatoncheires) against the Titans raged for ten years – an apocalypse that set forests alight and made the ocean boil, the land shudder and the sky crack... Anyone who has ever been caught out in a storm will understand how easy it would be in the days before our scientific understanding of nature to imagine the tumult in the skies as these warring gods!

THE WORK OF THE DEVIL

According to a French folktale, thunder was conjured up by the Devil who wanted to scare the wits out of humans by causing the sky to rumble and crack. He told God what he planned to do, but God went one better: he invented lightning to come before the thunder so that people could see his light and remember to put their trust in him.

Gods of thunder and lightning

Many cultures have a god whose specific role is to control thunder and lightning. Here are just a few:

Inanna She-who-must-be-obeyed! Sumerian sky goddess who had a dual role as life-nurturer and destroyer, presiding over both thunder and flood.

Lei Kung Chinese god of thunder.

Mamaragan Weather god of the Australian Aborigines, who rides a black cloud across the sky, hurling lightning bolts at trees and people. The thunder is the sound of his voice.

Thor Norse god with a magic hammer called Mjölnir. When he struck his hammer against stone, sparks flew and became lightning; when he threw it, it turned into a thunderbolt. He even has a day of the week named after him: Thursday.

Thunderbird Mythical bird of Native American culture. Thunder was the sound of the beating of his wings.

Zeus Leader of the Greek gods and compulsive philanderer, with a state-of-the-art pad on Mount Olympus. His armoury included the thunder and lightning bolt. To the Romans, he was known as Jupiter or Jove. Jehovah, the Hebrew god, was similarly free with his thunderbolts.

The wild hunt

Booming thunder, blinding lightning, swollen black clouds racing across the sky and general celestial mayhem? A storm, you think – but according to ancient German folklore, the ruckus is caused by a pack of wild hounds and thudding riders rampaging across the heavens. 'The raging host', or 'das wüthende Heer', was led by the god Woden (the Teutonic equivalent of the Norse Odin) – one of the earliest 'riders on the storm'.

English folklore had similar stories, as you can see in this poem by William Cowper:

> God moves in a mysterious way
> His wonders to perform;
> He plants his footsteps in the sea,
> And rides upon the storm.

From *Olney Hymns* by William Cowper (1731–1800)

The four quarters of the Earth

The ancient Greeks gave individual names to the winds. Obedient to their master, the god Aeolus, who lived on a floating island in the middle of sea, the chief ones were:

Boreas   The North Wind
Notus   The South Wind
Eurus   The East Wind
Zephyrus The West Wind

The rain-bringers

Chac Mayan rain god who also had responsibility for wind and lightning, and for the fertility of crops.

Indra Indian god of rain and thunder, and one of the chief deities of the Indian pantheon.

Leza High-level sky god of southern Africa whose chief role was that of rain-bringer. He might appear to humans as a vast cloud of dust.

Tlaloc Aztec rain god who inhabited a storm cloud and had multiple responsibilities – along with gentle rain, he also controlled drought and dispensed frost and lightning.

Toninlili The rain-bringer of the Navajo.

Ask...

In Native American rain-making ceremonies, an instrument called a bull-roarer was used to simulate the sound of thunder in the hope that – by means of sympathetic magic – the spirits would be tempted to imitate the sound and bring on a thunderstorm and rain.

... and ye shall receive

But you can sometimes overdo it. When, in June 2001, the inhabitants of Louisiana responded to Governor Mike Foster's call to pray for rain after weeks of drought, their efforts worked too well. The remnants of a tropical storm moved in from the Gulf of Mexico and deposited up to 1 metre (3 feet) of rain, causing extensive flooding. 'We may have prayed too long', said one resident.

Heavenly arc

In ancient times, the rainbow that may appear after a downpour was frequently seen as a bridge between Heaven and Earth. In Greece, it was sacred to the goddess Iris. In Scandinavia it was called 'Bifrost', the bridge of the gods, and also symbolized the celestial necklace belonging to the goddess Freya. In Biblical imagery, it was the sign given by God after the Deluge that he would never flood the world again.

Epilogue

✍

DESPITE ALL OUR scientific understanding of the weather, despite modern meteorology, central heating, air conditioning and snug, waterproof houses – despite all these advantages, we can still be overcome by the same awesome elemental forces as our ancestors. Unexpected and exceptionally heavy snowfall can cause traffic chaos, close schools and interrupt the normal running of everyday life. Tragically, hurricanes and floods can devastate whole areas and cause massive loss of life. Drought can cause crops to wither, cattle to die and people to starve, while vegetation, tinder-dry from lack of rain, can be all the fuel a wildfire needs to race across a landscape devouring all in its path, including human lives.

Equally the beauty of clear skies on a summer's day, or of a rainbow breaking through the clouds, or the perfection of an untouched blanket of snow on the ground on a winter's morning can overwhelm us in a very different way. Even with all our technological know-how, we cannot completely explain the awe-inspiring power of the weather.

However, by having a firm grasp of the basic principles of this book you will now be equipped with a solid understanding of weather formations; you will be able to try your hand at some forecasting of your own so you don't get caught out in a storm or in the freezing cold; and you will hopefully have an increased appreciation of and love for the mighty forces of the weather at play within our Earth's atmosphere.

Bibliography

Books

Dunlop, Storm, *Weather*. Collins, 2004.

Henson, Robert, *The Rough Guide to Weather*. Rough Guides, 2007.

Reynolds, Ross, *Guide to Weather*. Philip's, 2006.

Thunberg, Greta, *The Climate Book*. Allen Lane, 2022.

Woodward, Antony & Penn, Robb, *The Wrong Kind of Snow*. Hodder & Stoughton, 2008.

Websites

http://news.bbc.co.uk/weather/
The website of BBC (British Broadcasting Corporation) Weather.

http://www.metoffice.gov.uk/weather/
Website of the UK's official Meteorological Office.

http://www.weatheronline.co.uk/
Information on all aspects of the weather.

http://www.space.com/scienceastronomy/
Fascinating facts on climate and weather phenomena.

Index